Math Workbook Introduction

Wouldn't it be great to learn math by just watching television (or videos on YouTube)? Of course it would. However, the best method for learning a skill subject is by doing it. This is a workbook for learning math by doing it. This workbook for whole number operations contains the necessary definitions, rules, and procedures written with easy-to-understand terminology. Then, there are step-by-step examples, with the steps explained, in each section. After each example, it's Your Turn to do problems just like the example. This pattern continues in each section, step-by-step examples and more Your Turn problems. At the end of each section is a traditional homework section. All answers to Your Turn Problems and Homework are included. The workbook has enough space for you to write-in your solution steps and answers.

Supplements

Please visit the website "BobsMathClass.com." Available free to you are PowerPoint lessons and videos for each section. If you have any questions or comments concerning this workbook, website, or other workbooks, do not hesitate to email me.

robert.ramirez@vvc.edu

Bob Ramirez

Math is a Unique Subject.

Math is different than other subjects, you must do it. Practice is absolutely necessary. Just like a sport, you can not learn skills in a sport just by reading about it or watching it on T.V. Math is a skill subject and you must develop the skill. The development of any skill requires practice, practice, and even more practice. You must constantly review what you have learned. Math is also unique in the fact that it has a sequential learning pattern. There is an order as to how new concepts are taught. Each new concept and skill is based upon previously learned material. For example, if you do not know how to add, you will not be able to learn the multiplication facts. If you do not know how to multiply and subtract, there is no way you will learn how to divide. Each topic must be learned before you can go on to the next topic.

Math as a Foreign Language

It is very helpful to think of math as a foreign language. You will need to learn the meaning of its symbols. The numbers and symbols when put together are not unlike expressions and sentences in any language. It has its own vocabulary. Often the words used are just synonyms for commonly used words. Learn the terminology. If someone is speaking math without you knowing the terminology, it will just be impossible to comprehend. Because math is a language, you must practice speaking it; talk to yourself as you study and do homework. Learn vocabulary and definitions as you would learn words of any new language. Make note cards, word lists, and constantly review.

Keeping a Positive Attitude

Have you ever been frustrated or in a bad mood and you let your emotions get away from you? Maybe you might have said something or done something you wouldn't have done if you weren't in this state of mind? Many of us simply do not think clearly when we are upset. Keeping a positive attitude at math is extremely important. It is not easy to learn math if we are not thinking clearly. If you feel yourself getting frustrated, take a break. When you come back to it, review some previous material before going at it again. We need to do our best to stay in a calm and relaxed state of mind.

Math Anxiety

Math anxiety is defined as a feeling of tension and apprehension that interferes with one's ability to perform in mathematical situations. The symptoms may be physical or psychological such as nausea, sweating, increased blood pressure, memory loss, and loss of self-confidence. Because of this anxiety, normal functioning is decreased and skills necessary for learning and performing become more difficult. Math anxiety is usually caused by bad experiences in math situations from one's past. These experiences may be comments made by teachers or even parents. Math anxiety is a learned condition. Therefore, it can be unlearned by positive experiences and successes. Successes can be increased by developing good math study habits along with strategies for test preparation. Breathing exercises have shown to decrease anxiety. Try to reduce negative self-talk and replace it with positive self-talk. Keeping a positive attitude is essential to success in mathematics.

Essential Mathematics Table of Contents

Table of Contents Cont.

ISBN-13: 978-1499172065
ISBN-10: 1499172060
Printed April, 2014

1.1 Place Value of Whole Numbers

Whole Numbers: Whole numbers can be listed as 0,1,2,3,4,5,6,7,8,9,10,11, . . .

The ". . ." means that the pattern of numbers continues without end.

Natural Numbers: The natural numbers are the numbers 1, 2, 3, 4, 5, 6, . . .
The number 0, is not a natural number.

Infinity: Infinity is the concept that numbers continue to get larger and larger with no definite

end. There is no largest number.

Place Value:
The particular position of a digit within a number determines its place value. For example, it the
digit is 4 places from the left, the number is in the thousands place.

<pre>
                                                                   Hundred thousands
                  Hundred billions   Ten thousands
          Trillions   Ten billions   Hundred millions Millions   Thousands Hundreds   Tens   Ones
                                  Billions   Ten millions

          6 ,5   0   3 ,4   3   8,2   3   6,4   3   6
</pre>

Period: A period is a set of digits separated by a comma. Every three digits from the right is a
period. The period for the first three numbers is called the ones period. The period for the
second three numbers from the right is called the thousands period. The millions period in the
number above is the third period from the right: 438.

After the billions period is the trillions period. After the trillions period is the quadrillions
period. After the quadrillions period is the quintillions. These prefixes have a Latin origin. (bi-
2, tri-3, quad-4, quint-5, ...). Note: zillions and gazillions are made up words. They are not
actual periods in the whole number system.

When a whole number is written using the digits $0 \rightarrow 9$ it is said to be in the ***standard form***. The
position that each digit occupies determines its *place value*.

Essential Mathematics Workbook

Example 1. What does the digit 4 mean in each number?

   a. 23<u>4</u>,598   b. <u>4</u>56,901   c. 2<u>4</u>,355,567,222

Answers: a. 4 thousands   b. 4 hundred thousands   c. 4 billions

---

Your Turn Problem #1
What does the digit 7 mean in each number?

a. 573,289:_____

b. 213,570_____

c. 37,218,021,593_____

Writing Numbers in Words

Hyphens: When writing a number between 21 and 99, excluding 30, 40, 50, 60, 70, 80 and 90, a hyphen must be used between the two numbers.

Examples: 24 → twenty–four

73 → seventy–three

45 → forty–five   (40 is spelled forty, not fourty.)

Commas: Commas are used to separate each period. The comma is and must be used only after the words: thousand, million, billion, trillion, etc.

   Examples: twenty–three thousand, five hundred eighty–eight (23,588)

   five million, twenty–two thousand, eleven (5,022,011)

Note: For four-digit numbers, some authors choose not to use a comma to separate the thousands period from the ones period when the number is written with numbers. When the number is written with words, the comma is still used.

   Examples: 2010: two thousand, ten

   8156: eight thousand, one hundred fifty-six

And: The word "and" is only to be used with numbers that have decimals and mixed numbers. So for now, do not write the word "and".

Procedure: When asked to rewrite a number using words:

Proceeding left to right, write in words the number in each period, then write its period name (example: trillion, million, etc...), then write a comma, then proceed to the next period, etc...

Note: The only period which is not written is the ones period (the last period).

Example 2a.   Rewrite using words: 72,417

Answer: Seventy-two thousand, four hundred seventeen.

Example 2b.   Rewrite using words: 12,503,438,236,473

Answer: Twelve trillion, five hundred three billion, four hundred thirty-eight million, two hundred thirty-six thousand, four hundred seventy-three.

Example 2c.   Rewrite using words: 5,043,018

Answer: Five million, forty-three thousand, eighteen.

Your Turn Problem #2
Rewrite using words

a) 9,207:_____

b) 3,429,718:_____

Writing Words into Numbers

Procedure: When asked to rewrite a written number in words into numbers using digits:

Proceeding left to right, for each period (separated by commas), write the number using digits and place commas where they are placed in the written form using words.

If a period is not written, write "000", and continue to the next period.

Example 3a.   Rewrite using numbers:

Four billion, three hundred million, four hundred eight thousand, forty-six

Answer: 4,300,408,046

Example 3b.   Rewrite using numbers:

Fifty-six trillion, six million, eight

Answer: 56,000,006,000,008

Example 3c.   Rewrite using numbers:

Two hundred ninety-seven thousand, eight hundred

Answer: 297,800

Your Turn Problem #3

Rewrite using numbers:

a) One hundred thirty-five thousand, four hundred two:_____

b) Four billion, four hundred eight thousand, forty-six:_____

1.1 Homework: Place Value of Whole Numbers

Complete the place value system.

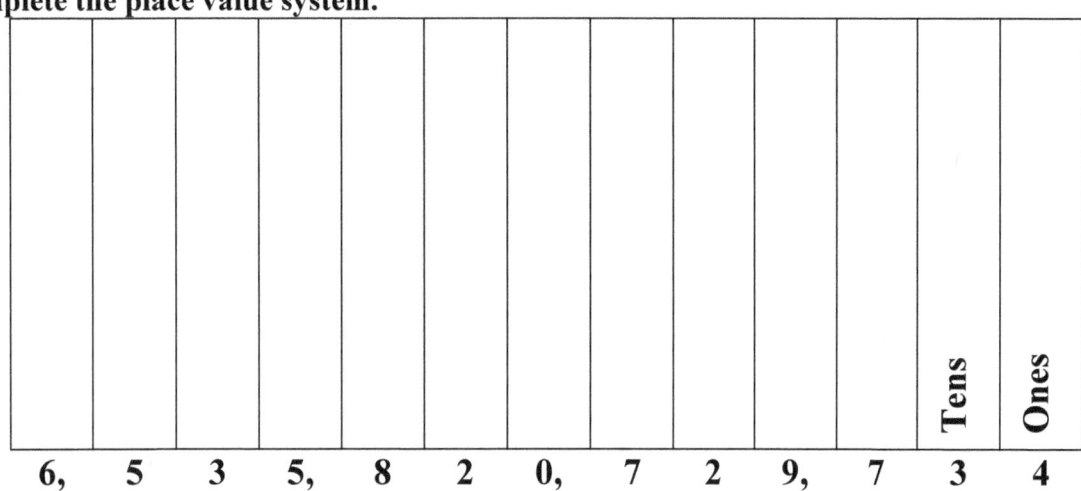

											Tens	Ones
6,	5	3	5,	8	2	0,	7	2	9,	7	3	4

Give the place values for the indicated digits.

1. 8 in the numeral 7,831

2. 5 in the numeral 27,573,429

3. 2 in the numeral 208,417,145

4. 4 in the numeral 347,123,831,566

5. 9 in the numeral 429,601

6. 0 in the numeral 8,675,309

Write each of the following in words.

7. 21:_____

8. 941:_____

9. 3,501:_____

10. 93,880:_____

1.1 Homework: Place Value of Whole Numbers cont.

Write each of the following in words.

11. 34,058,012:_____

12. 7,012,000,030,008:_____

Write each of the following numbers with digits instead of words.

13. Forty-seven:_____

14. Three hundred twenty-seven:_____

15. Five thousand, eighty-two:_____

16. One hundred thirty-five thousand, four hundred two: _____

17. Eight trillion, seven thousand: _____

18. Two million, four hundred thousand, eleven:_____

19. Forty-three thousand, nine hundred four:_____

20. Eight million, two hundred eighty:_____

1.2 Rounding and Ordering of Whole Numbers

Accuracy

Accuracy indicates how exact we need to be in expressing a value. For example if you need to know your approximate income at your job last year which was $32,882.46, you would probably answer "$33,000." Your accuracy was "to the nearest thousand."

Rounding Off to a Given Place Value:

Rounding off is the process of changing a number and expressing it to a certain accuracy. We will indicate accuracy by specifying a "given place value".

Procedure: When asked to round to a certain place value:
1. Underline the place value digit of desired accuracy.
2. Look at the digit to its right.
3. If that digit is 5 or larger, add one to the place value digit. If that digit is less than 5, leave the place value digit alone (do not change).
4. Rewrite all the digits to the right of the place value digit as zeros.

Example 1. Round the following to the nearest ten.

a) 68

    1. 6<u>8</u>

    2. Since the digit to the right is 5 or larger, add 1 to the tens place.

Answer: 70

b) 492

    1. 4<u>9</u>2

    2. Since the digit to the right is less than 5, leave the tens value digit alone.

Answer: 490

c) 497

    1. 4<u>9</u>7

    2. Since the digit to the right is 5 or larger, add 1 to the tens place.

Answer: 500

11

Your Turn Problem #1
Round the following to the nearest ten.

a) 34 _____   b) 857:_____   c) 6,796:_____

Example 2. Round the following to the nearest hundred.

a) 4,829   1. 4,829

         2. Since the digit to the right is less than 5, leave the hundreds digit alone.

Answer: 4,800

b) 7,951   1. 7,951

         2. Since the digit to the right is 5 or larger, add 1 to the hundreds.

Answer: 8,000 (79 + 1 = 80)

c) 72   1. 072

         2. Since the digit to the right is 5 or larger, add 1 to the hundreds.

Answer: 100

Your Turn Problem #2

Round the following to the nearest hundred.

a) 72,963 _____   b) 408,257 _____   b) 39 _____

Example 3. Round the following to the nearest thousand.

a) 76,856   1. 7<u>6</u>,856

                  2. Since the digit to the right is more than 5, add 1 to the thousands digit.

Answer: $\boxed{77,000}$

b) 39,476   1. 3<u>9</u>,476

                  2. Since the digit to the right is less than 5, leave the thousands digit alone.

Answer: $\boxed{39,000}$

c) 61,472,887   1. 61,47<u>2</u>,887

                     2. Since the digit to the right is more than 5, add 1 to the thousands digit.

Answer: $\boxed{61,473,000}$

Your Turn Problem #3

Round the following to the nearest thousand.

a) 45,378 _____   b) 425 _____

c) 789 _____

The process is the same for any place value, thousands, ten thousands, hundred thousands, millions, etc.

Example 4. Round the following to the nearest million.

a) 5,717,364   1. <u>5</u>,717,856

2. Since the digit to the right is more than 5, add 1 to the millions digit.

Answer: 6,000,000

b) 83,476,455   1. 8<u>3</u>,476,455

2. Since the digit to the right is less than 5, leave the millions digit alone.

Answer: 83,000,000

Your Turn Problem #4

Round the following to the nearest ten thousand.

a) 245,211 _____   b) 3,442,574 _____

Comparing Numbers

Whole numbers can be shown on a number line.

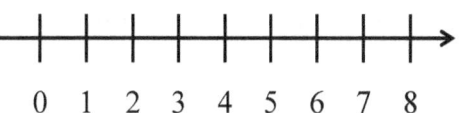

0 1 2 3 4 5 6 7 8

From the number line, we can see the order of numbers. For example, we can see that 2 is less than 7 because 2 is to the left of 7. For any two numbers on a number line, the number to the left is always the smaller number, and the number to the right is always the larger number.

We use the inequality symbols $<$ or $>$ to write the order of numbers.

Inequality Symbols

For any whole numbers a and b:

1. $a < b$ (read a is less than b) means a is to the left of b on the number line.
2. $a > b$ (read a is greater than b) means a is to the right of b on the number line.

    " $>$ " means greater than

    " $<$ " means less than

Note: The inequality symbol must always **point toward the smaller number**. You could also say it **opens up to the larger number**.

Examples: $16 > 5$ $3 < 8$

Example 5. Place the correct inequality symbol between the two numbers.   **Answers**

a) 15   12   a) $15 > 12$

b) 0   11   b) $0 < 11$

Your Turn Problem #5

Place the correct inequality symbol between the two numbers.

a)   34   17   b)   25   12   c)   400   760

1.2 Homework: Rounding and Ordering of Whole Numbers
Round each of the following numbers to the nearest ten

1. 346 _____

2. 3,512 _____

3. 13,515 _____

4. 2,397 _____

5. 4 _____

6. 438 _____

7. 76,429 _____

8. 571,597 _____

9. 111,302 _____

Round each of the following numbers to the nearest hundred

10. 746 _____

11. 3,551 _____

12. 13,961 _____

13. 29,952 _____

14. 712 _____

15. 39,748 _____

16. 711,881 _____

17. 42,973 _____

18. 63 _____

1.2 Homework: Rounding and Ordering of Whole Numbers cont.

Round each of the following to the nearest thousand

19. 7,346_____

20. 17,512_____

21. 19,836_____

22. 459_____

23. 501_____

24. 3,489_____

25. 773,412,521_____

26. 9,805_____

27. 7,000,824_____

Round each of the following to the indicated place value

28. 179,212 : ten thousand

Answer:_____

29. 45,873,712 : hundred thousand

Answer:_____

30. 226,771,333 : million

Answer:_____

31. 341,561,202 : ten million

Answer:_____

Place the correct inequality symbol between the two numbers. (< or >)

32. 15   30

33. 605   210

34. 5   10

35. 24   11

36. 93   0

37. 85   134

1.3 Addition of Whole Numbers

Addition of Whole Numbers and Properties under Addition.

If two numbers a and b are added, that operation can be expressed as $a + b = c$ where a and b are called *addends* and the result c is called the sum.

Addition is the operation where one amount is combined with another amount to get a total.

Example:
$$\begin{array}{r} 5 \\ + 2 \\ \hline 7 \end{array} \leftarrow \text{addends}$$

7 ⟵ Sum

Phrases for Addition

sum, total, increased by, more than, added to, plus

Example: The sum of 5 and 9 is 14.

Translates to: $5 + 9 = 14$

Example: "4 more than 10" translates to: $4 + 10$

or $10 + 4$

Notice that the order for addition does not matter. $4 + 10$ is the same as $10 + 4$.
This is called the Commutative Property of Addition.

Commutative Property of Addition

$a + b = b + a$ where a and b are numbers.

Example: $5 + 7 = 7 + 5$
$12 = 12$

Additive Identity Property

The number 0 is the additive identity because for any number a,

$a + 0 = a$ and $0 + a = a$.

The number 0 is the *identity element* of addition because when 0 is added to any number, the resulting sum of addition is the original number

Example: $5 + 0 = 5$

Parentheses

Parentheses: a mathematical expression which indicates: simplify inside first.

$($ $)$

Example: $5 + (7 + 3)$

$5 + 10$ Since there are parentheses, we will simplify inside the parentheses first.

15

Note: We would get the same answer if we added the 5 and 7 first. The order does not matter because the order in which the addition is performed will not matter.

Since the order does not matter; i.e., $5 + (7 + 3) = (5 + 7) + 3$ (both equal 15), this gives another property.

This is called the Associative Property of Addition.

Associative Property of Addition

$a + (b + c) = (a + b) + c$ where a, b and c are numbers.

Example: $4 + (9 + 5) = (4 + 9) + 5$

The numbers are in the same order. The parentheses are in different positions.

Your Turn Problem #1
State the indicated Property:

a) $12 + 0 = 12$ _____

b) $3 + 7 = 7 + 3$ _____

c) $3 + (5 + 8) = (3 + 5) + 8$ _____

Procedure: To solve an addition problem where the addends are written horizontally:

1. Rewrite the problem aligning digits by place value: one's place on right, ten's place second from right, etc...

2. Then add the columns, carrying as you go.

Example: Add: $8,456 + 2,484$

Rewrite the problem vertically

Add ones. We get 10 ones. Write the 0 in the ones column and 1 above the tens. This is called *carrying*.

Add tens. We get 14 tens. Write 4 in the tens column and 1 above the hundreds.

Add hundreds. We get 9 hundreds. Write 9 in the hundreds column.

Add thousands. We get 10 thousands

$$\begin{array}{r} {\scriptstyle 1\ \ 1} \\ 8\ 4\ 5\ 6 \\ +\ 2\ 4\ 8\ 4 \\ \hline 1\ 0\ 9\ 4\ 0 \end{array}$$

Answer: $\boxed{10{,}949}$

Your Turn Problem #2

Find the sum of 2,967 and 12,844.

Answer:_____

Key words such as "total" and "increased by," imply addition.

Adding with more than two addends.

If there are more than two addends, there are two different options as to how to proceed. Option one, write the addends vertically and add all addends at one time. Option two, add the first two addends. Then add the sum to the next addend. Continue until all of the addends are used.

Example: Add $54 + 64 + 98$

Option 1:
$$\begin{array}{r} 54 \\ 64 \\ +98 \\ \hline 216 \end{array}$$

Option 2: $54 + 64 + 98$

Add the first two addends.

$$\begin{array}{r} 54 \\ +64 \\ \hline 118 \end{array}$$

$118 + 98$

$$\begin{array}{r} 118 \\ +98 \\ \hline 216 \end{array}$$

Example 3. The yearly profit for DR Construction was $78,216 in 2004, $153,917 in 2005, and $85,098 in 2006. What is the total profit for these three years?

Solution: The key word in this problem is "total".

$$\begin{array}{r} 78216 \\ 153917 \\ + 85098 \\ \hline 317231 \end{array}$$

Answer: The total profit for the three years was $317,231.

Your Turn Problem #3

Laura's yearly salary is $58,500. Next year, her salary will increase by $7,000. What will her yearly salary be next year?

Answer:_____

1.3 Homework: Addition of Whole Numbers

Name the property that is illustrated.

1. $5 + 6 = 6 + 5$: _____

2. $3 + (9 + 12) = (3 + 9) + 12$: _____

3. $4 + (5 + 6) = (4 + 5) + 6$: _____

4. $7 + 0 = 7$: _____

5. $(11 + 4) + 5 = 11 + (4 + 5)$: _____

6. $11 + 3 = 3 + 11$: _____

Add the following.

7. $12 + 5$

8. $17 + 9$

9. $23 + 16$

10. $33 + 27$

11. $78 + 35$

12. $96 + 47$

13. $30 + 49$

14. $93 + 89$

1.3 Homework: Addition of Whole Numbers cont.

Add the following.

15. 321+156

16. 872+731

17. 507+299

18. 857+765

19. 941 + 873

20. 3,501 + 2,916

21. 93,880 + 45,371

22. 1,342 + 978 + 128

23. 85,864 + 72,586

24. 542,682 + 97,388

25. 727 + 4,871 + 609

26. 855 + 49,444 + 7,260

1.3 Homework: Addition of Whole Numbers cont.

Add the following.

27. Find the total of 23 and 17.

28. Find the sum of 867 and 5,309.

29. Find a number that is 18 more than 154.

30. What is the sum of 316 and 400?

31. A bowler scored 201, 157, and 198 in three games. What was the total score for those games?

32. Driving to California, Allison drove 871 miles on Monday, 612 miles on Tuesday, and 977 miles on Wednesday. How many miles did she drive on the three days?

33. Corrine spent $364 for tuition, $583 for books, and $35 for parking during one semester. What was the total cost for tuition, books, and parking for that semester?

1.4 Subtraction of Whole Numbers

Subtraction can be expressed by the equation $a - b = c$, where a is the *minuend*, b is the *subtrahend*, and c is the *difference*. $a - b = c$ is only true if $c + b = a$ is true.

Subtraction is the operation where one amount is "taken away" from another amount leaving the difference.

Example: 18 ⟵ minuend
          − 5 ⟵ subtrahend
          ————
           13 ⟵ difference

Phrases that indicate Subtraction:

Minus, Subtract, Take Away, Difference, **Subtracted from * , Less than *,** Decreased by

Example 1:	Translated
12 minus 4	12 − 4
The difference of 9 and 2	9 − 2
5 subtracted from 17	17 − 5 *
4 less than 10	10 − 4 *

Note: The order for the phrases with * are reversed

In this section, we are only working with whole numbers, i.e., 0, 1, 2, 3 … We are not working with negative numbers, but we understand a little about them. If the temperature is −17°, it's cold. If your checking account is $−24, you're in the negative.

So, 4 less than 10 translates to 10 − 4. If you did it incorrectly, and wrote 4 − 10, the answer would be −6. Subtraction is not commutative; order does matter.

Your Turn Problem #1
Translate and simplify.

a) 15 subtracted from 49.

b) The difference of 8 and 5.

Answer:_____

Answer:_____

Borrowing

Borrowing is the process that is used when the lower digit is larger than the upper digit.

Procedure for Subtraction - Borrowing:

1. Line out the digit to the left of the upper digit, subtract 1 from it and write this new digit above.

2. Add 10 to the original upper digit by writing a 1 in front of it.

3. Subtract the lower digit.

4. Proceed to the next column on the left.

Example 2a:

$$
\begin{array}{r}
5\ 7\ 2 \\
-\ 4\ 4\ 7 \\
\end{array}
$$

Solution:

$$
\begin{array}{r}
5\ \overset{6}{\cancel{7}}\ \overset{1}{2} \\
-\ 4\ 4\ 7 \\
\hline
1\ 2\ 5 \\
\end{array}
$$

Answer: | 125 |

Example 2b:

$$853 - 368$$

Solution:

$$
\begin{array}{r}
\overset{7}{\cancel{8}}\ \overset{14}{\cancel{5}}\ \overset{1}{3} \\
-\ 3\ 6\ 8 \\
\hline
4\ 8\ 5 \\
\end{array}
$$

Answer: | 485 |

Borrowing from 0.

When the digit(s) to the left of the upper digit is 0, the 0(s) are lined out and replaced by 9(s), and then the first non zero digit to the left is borrowed from.

Example 2c:

$$\begin{array}{r} 4\ 0\ 0\ 2 \\ -\ 2\ 3\ 7\ 5 \\ \hline \end{array}$$

Since 5 is greater than 2, borrow a "1" from the 4 and write 9's above the zeros up to the last number. The last number gets a 1 in front of it.

$$\begin{array}{r} 3\ 9\ 9 \\ \cancel{4}\ 0\ 0\ {}^{1}2 \\ -\ 2\ 3\ 7\ 5 \\ \hline 1\ 6\ 2\ 7 \end{array}$$

Your Turn Problem #2

Subtract the following:

a) 786 – 528

b) 45,354 – 29,676

c) 98,000 – 17,827

a) Answer:_____

b) Answer:_____

c) Answer:_____

Example 3: In June the Big Bear Boutique sold $24,760 worth of merchandise, but in July, it sold only $19,458 worth of merchandise. How much more did the boutique sell in June than in July?

Solution: "How much more" indicates subtraction.

24,760 – 19,458

= 5,302

Answer: The boutique sold $5,302 more in June than July.

Your Turn Problem #3

The attendance for a concert was 12,329 on Friday and 23,421 on Saturday. How many more people attended on Saturday than on Friday?

Answer: _____ more people attended on Saturday than on Friday.

Subtracting with three or more numbers.

If there are more than two numbers, subtract using the first two numbers. Then subtract using the result from the first two numbers and the third number. Continue until all of the numbers are used.

Example 4. Perform the indicated subtraction: $186 - 54 - 61$

Solution: Subtract using the first two numbers.

$$\begin{array}{r} 186 \\ -54 \\ \hline 132 \end{array}$$

Subtract using the result from the first two numbers and the third number.

$$132 - 61 \qquad \begin{array}{r} 132 \\ -61 \\ \hline 71 \end{array}$$

Answer: $\boxed{71}$

Your Turn Problem #4

Perform the indicated subtraction.

$$543 - 89 - 77$$

Answer:_____

1.4 Homework: Subtraction of Whole Numbers

Perform the indicated subtraction.

1. $25 - 12$

2. $95 - 22$

3. $73 - 45$

4. $82 - 24$

5. $34 - 28$

6. $77 - 39$

7. $43 - 29$

8. $38 - 19$

9. $63 - 28$

10. $142 - 98$

1.4 Homework: Subtraction of Whole Numbers cont.

Perform the indicated subtraction.

11. $781 - 362$ **12.** $543 - 326$

13. $80 - 34$ **14.** $70 - 42$

15. $100 - 73$ **16.** $400 - 156$

17. $900 - 713$ **18.** $500 - 242$

19. $900 - 537$ **20.** $600 - 492$

21. $4,000 - 2,254$ **22.** $8,000 - 5,317$

1.4 Homework: Subtraction of Whole Numbers cont.

Perform the indicated subtraction.

23. $12,000 - 6,337$

24. $14,000 - 8,773$

25. $346 - 279$

26. $3,512 - 2,863$

27. $13,521 - 5,931$

28. $2,000 - 741$

29. $15,000 - 9,215$

30. $970 - 548$

31. $1,208 - 653$

32. $6,200 - 4,737$

33. $90,000 - 15,605$

34. $16 - 5 - 8$

1.4 Homework: Subtraction of Whole Numbers cont.

Perform the indicated subtraction.

35. $34 - 15 - 9$

36. $78 - 23 - 17$

37. $512 - 73 - 87$

38. $900 - 156 - 98$

39. $834 - 93 - 185$

40. Subtract 23 from 47.

41. Subtract 965 from 14,811.

42. Find the difference of 24 and 17.

43. Find the difference of 76 and 24.

1.4 Homework: Subtraction of Whole Numbers cont.
Solve the following applications.

44. David's monthly pay of $2600 was decreased by $475 for taxes. What amount of pay did he receive after taxes?

45. Henry has $728 in cash and wants to buy a laptop that costs $1,241. How much more money does he need?

46. Brian's checking account has a balance of $575. Brian wrote three checks for $54, $37, and $143. What was the new balance in the checking account?

47. To earn an A in a math class, Patricia needs to have 400 points. If she has already earned 317 points in the class, how many more does she need to earn an A in the class?

Practice Test 1

Rewrite the following using words.

1. 9,742,013: _____

2. 5,023,960,227:_____

Name the property that is illustrated

3. $7 + 0 = 7$:_____

4. $3 + (8 + 2) = (3 + 8) + 2$:_____

5. $12 + 5 = 5 + 12$:_____

Perform the indicated operation.

6. $58 + 896$

7. $62 + 797 + 518$

8. $8,959 + 7,051 + 857$

9. $5,695 + 82,517$

10. $492 + 29,899 + 97$

11. $67 + 429 + 13 + 515$

Practice Test 1 cont.

12. What is the sum of 781 and 691?

13. What is the total of 35, 628, and 97?

Round to the indicated place value.

14. 44,458 ; thousand

15. 49,845 ; thousand

16. 9,375 ; hundred

17. 239 ; hundred

Perform the indicated operation.

18. $940 - 156$

19. $37,252 - 6,278$

20. $27,000 - 5,765$

21. $9,000 - 4,275$

Practice Test 1 cont.

22. $5,486 - 2,150 - 859$

23. $42,000 - 19,253 - 8,158$

24. Find the difference of 512 and 257.

25. Subtract 396 from 732.

Use the symbol < or > to complete the statements.

26. 0 _____ 19

27. 76 _____ 23

28. 2,009 _____ 989

29. 319 _____ 725

30. Victor Valley College has 11,531 students and Crafton Hills College has 9,697 students. How many more students does Victor Valley College have?

31. Martin has $500 to spend on school costs. He spends $234 for tuition and $158 for books. How much money does he have left to spend on other school costs?

32. Jaclyn read 184 pages on Monday, 314 pages on Tuesday, and 236 pages on Wednesday. What is the total number of pages she read for the three days?

2.1 Multiplication of Whole Numbers

Multiplication is the process that shortens **repeated addition** with the same number.

For example, $8 + 8 + 8 + 8 + 8 = 5 \cdot 8 = 40$ (Multiplication symbol: $\times$ or $\cdot$)

Multiplication can also be indicated with parentheses: $5(8) = 40$ or $(5)(8) = 40$.

The multiplication of two numbers a and b can be expressed by the equation $a \cdot b = c$ where a and b are *factors* and c is called the *product* (a and b are also called multiplicands).

Example:

$$\begin{array}{r} 4 \\ \times 3 \\ \hline 12 \end{array}$$

$\longleftarrow$ Factors (or multiplicands)

$\longleftarrow$ Product

Notice that the order for multiplication does not matter. 4×10 is the same as 10×4.

This is called the Commutative Property of Multiplication.

Commutative Property of Multiplication

$a \cdot b = b \cdot a$ where a and b are numbers.

Example: $5 \cdot 7 = 7 \cdot 5,$
$\qquad\quad 35 = 35$

Multiplicative Identity Property

The number 1 is the identity of multiplication because for any number a, $a \cdot 1 = a$ and $1 \cdot a = a$.

The number 1 is the identity element of multiplication because when 1 is multiplied by any number, the resulting product is the original number.

Example: $5 \cdot 1 = 5$

Associative Property of Multiplication

$a \cdot (b \cdot c) = (a \cdot b) \cdot c$ where a, b, and c are numbers.

Example: $(9 \cdot 5) \cdot 7 = 9 \cdot (5 \cdot 7)$

The numbers are in the same order. The parentheses are in different positions.

Multiplication Property of Zero

Multiplying any whole number by 0 gives the product 0.

Example: $5 \cdot 0 = 0$

Distributive Property of Multiplication over Addition

For any whole numbers a, b, and c, $a \cdot (b + c) = a \cdot b + a \cdot c$

Note: Often, this property is just called the *Distributive Property*

Example: $5 \cdot (7 + 3) = 5 \cdot 7 + 5 \cdot 3$ (Verify the result is 50 on both sides.)

Your Turn Problem #1:

State the indicated Property.   Answers

a) $12 \cdot 1 = 12$ a) _____

b) $3 \cdot 7 = 7 \cdot 3$ b) _____

c) $3 \cdot (5 + 2) = 3 \cdot 5 + 3 \cdot 2$ c) _____

d) $4 \cdot 0 = 0$ d) _____

e) $4 \cdot (5 \cdot 3) = (4 \cdot 5) \cdot 3$ e) _____

Multiplication Facts

Even though we live in a society where calculators are cheap and readily available, to be competent and confident in mathematics, you should have memorized the multiplication facts for one digit numbers.

Multiplication Table

	1	2	3	4	5	6	7	8	9	10	11	12
1	1	2	3	4	5	6	7	8	9	10	11	12
2	2	4	6	8	10	12	14	16	18	20	22	24
3	3	6	9	12	15	18	21	24	27	30	33	36
4	4	8	12	16	20	24	28	32	36	40	44	48
5	5	10	15	20	25	30	35	40	45	50	55	60
6	6	12	18	24	30	36	42	48	54	60	66	72
7	7	14	21	28	35	42	49	56	63	70	77	84
8	8	16	24	32	40	48	56	64	72	80	88	96
9	9	18	27	36	45	54	63	72	81	90	99	108
10	10	20	30	40	50	60	70	80	90	100	110	120
11	11	22	33	44	55	66	77	88	99	110	121	132
12	12	24	36	48	60	72	84	96	108	120	132	144

1.5 Practice Set

1. $5 \times 3 = $ ___15___

2. $7 \times 4 = $ ___28___

3. $4 \times 9 = $ _____

4. $7 \times 8 = $ _____

5. $9 \times 9 = $ _____

6. $7 \times 7 = $ _____

7. $8 \times 4 = $ _____

8. $5 \times 7 = $ _____

9. $4 \times 8 = $ _____

10. $8 \times 9 = $ _____

11. $7 \times 3 = $ _____

12. $7 \times 5 = $ _____

13. $6 \times 6 = $ _____

14. $4 \times 6 = $ _____

15. $8 \times 3 = $ _____

Multiplying a multi-digit number by a single-digit number.

1.	Rewrite the product vertically aligning the ones digit from each factor. Write the multi-digit factor on top and the single-digit factor on the bottom.

Examples:
$$\begin{array}{r} 423 \\ \times 8 \\ \hline \end{array} \qquad \begin{array}{r} 578 \\ \times 9 \\ \hline \end{array}$$

2. Then, starting from the last digit of the top factor, multiply each digit by the bottom factor. Write the *ones* digit of each product below the line. If there is a *tens* digit, carry it. Add it to the next product on the left.

3. The result of the last product will be written in front with no carrying involved.

Example 2a. Multiply 75×9.

Solution: Write vertically. Then multiply, carrying as you go from right to left.

$$\begin{array}{r} 75 \\ \times\ 9 \\ \hline \end{array} \rightarrow \begin{array}{r} 4 \\ 75 \\ \times\ 9 \\ \hline 5 \end{array} \rightarrow \begin{array}{r} 4 \\ 75 \\ \times\ 9 \\ \hline 675 \end{array}$$

$$(5\times9 = 45) \qquad (7\times9 = 63, \\ 63+4 = 67)$$

Answer: $\boxed{675}$

Example 2b. Multiply 423×8.

Solution: Write vertically. Then multiply, carrying as you go from right to left.

$$\begin{array}{r} 423 \\ \times\ 8 \\ \hline \end{array} \rightarrow \begin{array}{r} 2 \\ 423 \\ \times\ 8 \\ \hline 4 \end{array} \rightarrow \begin{array}{r} 1\ 2 \\ 423 \\ \times\ 8 \\ \hline 84 \end{array} \rightarrow \begin{array}{r} 1\ 2 \\ 423 \\ \times\ 8 \\ \hline 3384 \end{array}$$

$$(3\times8 = 24) \qquad (2\times8 = 16, \qquad (4\times8 = 32, \\ 16+2 = 18) \qquad 32+1 = 33)$$

Answer: $\boxed{3,384}$

Your Turn Problem #2
Multiply the following.

a) 24×9 b) 347×6

Answer:_____   Answer:_____

Multiplication of Multiple Digit Factors

Procedure: If presented with a problem where both factors have more than one digit:

1.   Beginning with the one's place digit of the factor written below, multiply by all the digits of the factor written above (carrying as you go) writing the product directly below (called the partial product).

2.   Next with the ten's place digit of the factor below, multiply it by all the digits written above (carrying as you go). Place a zero to the right and then write the partial product.

3.   Next with the hundred's place digit of the factor below, multiply it by all the digits written above (carrying as you go). Place two zeros to the right and then write the partial product.

4.   Continue this process, increasing the number of zeros by one, until all the digits in the factor below have been multiplied.

5.   Add the partial products for the product to the problem.

Example 3a. Multiply 38×27.

Solution: Write vertically. Then multiply, carrying as you go from right to left.

$$
\begin{array}{r} 38 \\ \times\ 27 \\ \hline \end{array}
\rightarrow
\begin{array}{r} \overset{5}{} \\ 38 \\ \times\ 27 \\ \hline 266 \end{array}
\rightarrow
\begin{array}{r} 38 \\ \times\ 27 \\ \hline 266 \\ 0 \end{array}
\rightarrow
\begin{array}{r} \overset{1}{} \\ 38 \\ \times\ 27 \\ \hline 266 \\ 760 \end{array}
\rightarrow
\begin{array}{r} 38 \\ \times\ 27 \\ \hline 266 \\ +760 \\ \hline 1026 \end{array}
$$

$$(38\times7=266) \qquad\qquad (38\times2=76)$$

Answer: $\boxed{1,026}$

Note: The zero is added after the 76 because the 2 in the bottom factor represents 20 since it is in the tens place.

Example 3b. Multiply 472×136.

Solution: Write vertically. Then multiply, carrying as you go from right to left.

$$
\begin{array}{r} 472 \\ \times\ 236 \\ \hline \end{array}
\rightarrow
\begin{array}{r} 472 \\ \times\ 236 \\ \hline 2832 \end{array}
\rightarrow
\begin{array}{r} 472 \\ \times\ 236 \\ \hline 2832 \\ 14160 \end{array}
\rightarrow
\begin{array}{r} 472 \\ \times\ 236 \\ \hline 2832 \\ 14160 \\ 94400 \end{array}
\rightarrow
\begin{array}{r} 472 \\ \times\ 236 \\ \hline 2832 \\ 14160 \\ +94400 \\ \hline 111392 \end{array}
$$

$$(472\times6=2832) \qquad (472\times3=1416) \qquad (472\times2=944)$$

Answer: $\boxed{111,392}$

Note: The two zeros are added after the 944 because the 2 in the bottom factor represents 200 since it is in the hundreds place.

Your Turn Problem #3

Multiply the following.

a) 45×28

b) 634×32

Answer:_____   Answer:_____

Multiplying with three factors.

Procedure: If presented a problem where there factors are written horizontally:
1. Rewrite the first two and find their product.
2   Second, multiply their product by the third factor. Continue this process until all the factors have been multiplied.

Example 4. Multiply: $24 \times 8 \times 17$.

Solution:

First, find the product of 24 and 8.   Second, multiply this result with the next factor, 17.

$$
\begin{array}{r}
24 \\
\times\, 8 \\
\hline
192
\end{array}
\qquad\qquad
\begin{array}{r}
192 \\
\times\, 17 \\
\hline
1344 \\
+1920 \\
\hline
3264
\end{array}
$$

Answer: $\boxed{3264}$

Your Turn Problem #4
Multiply the following.
a) $12 \times 7 \times 5$ b) $24 \times 15 \times 37$
Answer:_____ Answer:_____

Multiplying by multiples of 10.

Example 5a. Multiply: 40×30

        Solution:

$$
\begin{array}{r}
40 \\
\times\,30 \\
\hline
00 \\
+1200 \\
\hline
\end{array}
$$

        Answer: 1200

Notice we can take a shortcut by multiplying the 4 and 3 and then writing the sum of the number of zeros at the end of each number being multiplied. There is one zero after the 4. There is one zero after the 3. The sum of zeros after each number is two. Therefore, after we find the product of 4 and 3, we can write two zeros at the end for our answer.

Example 5b. Multiply: 600×40.

Solution: $6 \times 4 = 24$. The sum of zeros after each number is three. So, we can add three zeros to the end of the product of 6 and 4.

Answer: $\boxed{24,000}$

Your Turn Problem #5
Multiply the following.

a) 900×70 b) 3500×400

Answer:_____   Answer:_____

2.1 Homework: Multiplication of Whole Numbers

Name the property that is being illustrated. (Do not evaluate, just name the property)

1. $5 \cdot 8 = 8 \cdot 5$: _____

2. $8 \cdot 1 = 8$: _____

3. $5 \cdot 0 = 0$: _____

4. $7 \cdot 6 = 6 \cdot 7$: _____

5. $8 \cdot (3 \cdot 5) = (8 \cdot 3) \cdot 5$: _____

6. $0 \cdot 8 = 0$: _____

7. $2 \cdot (4+9) = (2 \cdot 4) + (2 \cdot 9)$: _____

8. $1 \cdot 5 = 5$: _____

9. $4 \cdot (3 \cdot 5) = (4 \cdot 3) \cdot 5$: _____

Perform the multiplication.

10. 12×8

11. 15×7

12. 24×8

13. 36×7

2.1 Homework: Multiplication of Whole Numbers cont.

Perform the multiplication.

14. 67×23

15. 68×36

16. 95×64

17. 48×39

18. 46×38

19. 62×11

20. 104×17

21. 346×24

22. 652×18

23. 521×78

2.1 Homework: Multiplication of Whole Numbers cont.

Perform the multiplication.

24. 234×112

25. 206×108

26. 612×430

27. 450×210

28. 616×911

29. 592×300

30. $6{,}218 \times 98$

31. $3{,}217 \times 445$

2.1 Homework: Multiplication of Whole Numbers cont.

Perform the multiplication.

32. $4,121 \times 342$

33. $13 \times 16 \times 5$

34. $127 \times 34 \times 62$

35. $128 \times 216 \times 52$

36. $5,000 \times 300$

37. $3,600 \times 200$

38. $7,000 \times 40$

39. $12,000 \times 700$

2.1 Homework: Multiplication of Whole Numbers cont.

40. Find the product of 12 and 7

41. Find the product of 42 and 512

Solve the following applications.

42. One tablespoon of olive oil contains 125 calories. How many calories are in 3 tablespoons of olive oil?

43. The textbook for a course in history costs $54. There are 35 students in the class. Find the total cost of the history books for the class.

44. The seats in the mathematics lecture hall are arranged in 12 rows with 34 seats in each row. Find how many seats are in this room.

2.1 Homework: Multiplication of Whole Numbers cont.

Solve the following applications.

45. An apartment building has three floors. Each floor has five rows of apartments with four apartments in each row. How many apartments are in the building?

46. A koi farm has 7 fish ponds. If each pond has 2,000 koi, how many koi does the farm have?

47. Each classroom in the Technology Center has 36 computers. If the Technology Center has 12 classrooms, how many computers does the Technology Center have in its classrooms?

48. A truck holds 12 gallons of gas. If the truck gets 20 miles per gallon, how far can the truck go on a full tank of gas?

2.2 Division of Whole Numbers

Division defines the operation of finding how many groups of a certain number (the divisor) are contained in another number or amount (the dividend). The answer is the quotient.

Terminology

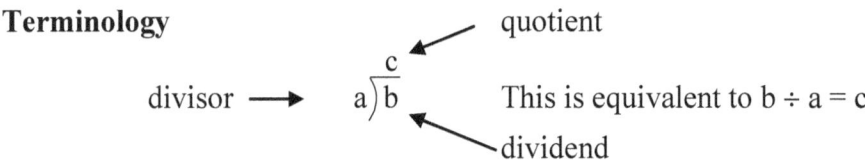

The following are equivalent forms of division.

$$12 \div 3, \qquad 3\overline{)12}$$

Both of these division problems equal 4. (12 is the dividend, 3 is the divisor, and 4 is the quotient.)

Related Multiplication Sentence

The division problem of $20 \div 5$ is defined to be the number that when multiplied by 5 gives 20.

$$20 \div 5 = 4 \quad \text{related multiplication sentence: } 5 \times 4 = 20$$

(The last two numbers, 5 and 4, multiply to equal the first number, 20.)

Write a related multiplication sentence to: $63 \div 9 = 7$

$$\boxed{} \times \boxed{} = \boxed{}$$

Division by 1

Any number divided by 1 is that same number:

$$a \div 1 = a \quad , \quad 1\overline{)a}^{\,a}$$

Example 1: $12 \div 1 = 12 \quad , \quad 1\overline{)12}^{\,12}$

Your Turn Problem #1

Divide the following.

a) $21 \div 1$ b) $34 \div 1$ c) $1\overline{)85}$ d) $1\overline{)7}$

Answer:_____   Answer:_____   Answer:_____   Answer:_____

Dividing a number by itself

Any number divided by itself is 1.

$$a \div a = 1 \quad , \quad a\overline{)\overset{1}{a}}$$

Example 2: $12 \div 12 = 1 \quad , \quad 12\overline{)\overset{1}{12}}$

Your Turn Problem #2

Divide the following.

a) $11 \div 11$ b) $28 \div 28$ c) $5\overline{)5}$ d) $15\overline{)15}$

Answer:_____   Answer:_____   Answer:_____   Answer:_____

Dividing into zero (Zero is the dividend.)

Zero divided by any nonzero number is zero.

$$0 \div a = 0 \quad , \quad a\overline{)\overset{0}{0}} \quad \text{where 'a' is not equal to zero.}$$

Recall: $\text{divisor}\overline{)\overset{\text{quotient}}{\text{dividend}}} \Rightarrow \text{dividend} \div \text{divisor} = \text{quotient}.$

If the dividend is 0, and the divisor (the other number) is not zero, the answer is zero.

Example 3: $0 \div 12 = 0 \quad , \quad 12\overline{)\overset{0}{0}}$

(Note: The last two numbers, 12 and 0, multiply to equal the first number, 0.)

Your Turn Problem #3

Divide the following if possible.

a) $0 \div 8$ b) $0 \div 15$ c) $7\overline{)0}$ d) $21\overline{)0}$

Answer:_____   Answer:_____   Answer:_____   Answer:_____

Division by zero (Zero is the divisor.)

A nonzero number divided zero is **undefined** (Not Zero!).

$a \div 0 :$ undefined where 'a' does not equal zero.

$0\overline{)12} :$ *undefined*

Example 4: $12 \div 0 :$ undefined , $0\overline{)12} :$ undefined

$12 \div 0 = 0$ would be incorrect.

(The last two numbers, 0 and 0, **do not** multiply to equal the first number, 12.)

Your Turn Problem #4

Divide the following if possible.

a) $9 \div 0$ b) $13 \div 0$ c) $0\overline{)4}$ d) $0\overline{)81}$

Answer:_____   Answer:_____   Answer:_____   Answer:_____

Restating:

If zero is before the division symbol, the answer is zero.

If zero is after the division symbol, the answer is undefined.

If zero is inside the division box, the answer is zero.

If zero is in front of the division box, the answer is undefined.

Zero divided by zero

When zero is divided by zero, it is called *indeterminate*.

$$0 \div 0 \text{ is indeterminate }, \quad 0\overline{)0} : \text{ indeterminate}$$

The reason $0 \div 0$ is indeterminate is because we can actually write any number and the related multiplication sentence would be true.

Example 5: Choose any number to write after the "=" sign. $0 \div 0 = 12$

The related multiplication sentence would be $0 \times 12 = 0$ which is true. So we could write any number and the related multiplication statement would be true.

Therefore, we must answer the problem of $0 \div 0$ as indeterminate.

Your Turn Problem #5

Divide the following if possible.

a) $0 \div 0$ b) $0\overline{)0}$

a)   _____   b)   _____

Please note: Some students have a tendency to draw a line through zeros, $\varnothing$. This symbol has a meaning to it. It is "no solution" or "empty set". Please do not use this symbol right now. It is used to indicate that there is no solution to an equation. We are not solving equations right now. So if the answer is undefined, you must write undefined. If the answer is indeterminate, you must write indeterminate. If the answer is zero, you must write 0.

Example 6. Divide: $4,224 \div 12$

<u>Long Division Procedure:</u>

1. Write the problem in long division format.

    If problem is given in this form "*dividend ÷ divisor*", change to: $divisor\overline{)dividend}$.

    $4,224 \div 12$ change to $12\overline{)4224}$

2. Determine the placement of the first digit on left in the quotient by determining how many digits of the dividend are necessary to accommodate the divisor.

    Take the first digit on the left in the dividend "4".

    Now ask, will $12\overline{)4}$?   Answer: No.

    Now take the first two digits on left in the dividend "42". Ask, will $12\overline{)42}$?

    Answer: Yes

    Therefore, the first digit on the left in the quotient is directly above the last digit of 42.

    $$12\overline{)4\overset{\downarrow}{2}24}\,.$$

3. Determine how many groups of the divisor are contained in the number obtained in Step 2. How many 12's are contained in 42? We want to find "what number multiplied by 12 will give us a result as close to 42 without going over".

    $12 \times 4 = 48$ (too big); $12 \times 3 = 36$

    Answer is 3. $12\overset{\ \ 3}{\overline{)4224}}$

4. Multiply the digit obtained from the quotient in step 3 by the divisor. Place that product below the dividend with the digit on the right directly below the digit in the quotient.

    $12 \times 3 = 36$

    $$\begin{array}{r} 3 \\ 12\overline{)4224} \\ 36 \end{array}$$

5. Subtract product from 42 and then bring down next digit from dividend on right (use an arrow)

$$
\begin{array}{r}
3 \\
12\overline{)4224} \\
-36\!\downarrow \\
\hline
62
\end{array}
$$

6. Go back to Step 3, and repeat the process using the bottom line as the new dividend.

How many "12's" in 62?

Answer is 5; $12 \times 5 = 60$.

Place 5 in the quotient, then repeat steps 4 and 5.

$$
\begin{array}{r}
35 \\
12\overline{)4224} \\
-36 \\
\hline
62 \\
-60\!\downarrow \\
\hline
24
\end{array}
$$

7. Repeat the process using the bottom line as the new dividend.

How many "12's in 24?

Answer is 2; $12 \times 2 = 24$.

Place 2 in the quotient, then continue steps 4 and 5.

$$
\begin{array}{r}
352 \\
12\overline{)4224} \\
-36 \\
\hline
62 \\
-60 \\
\hline
24 \\
-24 \\
\hline
0
\end{array}
$$

Answer: $\boxed{352}$

You can check by multiplying the divisor and the quotient. This will give the dividend.

$12 \times 352 = 4224$

Useful Technique for Long Division

When starting a long division problem, it may be beneficial to just write out the products of the divisor and numbers 1 through 9.

Example 6 (again): $4224 \div 12$

1. First rewrite in long division format.

    $12 \overline{)4224}$

2. Write the products of the divisor, 12, and the numbers 1-9 (shown to the right).

3. Determine where the first number will be placed in the quotient (above the division box). 12 will not divide into 4 but it will divide into 42. Therefore the first digit of the quotient is written above the 2.

4. Looking at our products to the right, we want to find the product closest to 42 without going over. Answer: 3.
    Write the 3 in the quotient, multiply, then subtract and bring down the next digit.

    $$\begin{array}{r} 3 \\ 12\overline{)4224} \\ -36 \\ \hline 62 \end{array}$$

5. Do it again. Find the product to the right closest to 62 without going over. Answer:5

    $$\begin{array}{r} 35 \\ 12\overline{)4224} \\ -36 \\ \hline 62 \\ -60 \\ \hline 24 \end{array}$$

6. Do it again. Find the product to the right closest to 24 without going over. Answer:2

    $$\begin{array}{r} 352 \\ 12\overline{)4224} \\ -36 \\ \hline 62 \\ -60 \\ \hline 24 \\ -24 \\ \hline 0 \end{array}$$

Answer: $\boxed{352}$

$\begin{array}{r}12\\ \times 1\\ \hline 12\end{array}$	$\begin{array}{r}12\\ \times 2\\ \hline 24\end{array}$	$\begin{array}{r}12\\ \times 3\\ \hline 36\end{array}$
$\begin{array}{r}12\\ \times 4\\ \hline 48\end{array}$	$\begin{array}{r}12\\ \times 5\\ \hline 60\end{array}$	$\begin{array}{r}12\\ \times 6\\ \hline 72\end{array}$
$\begin{array}{r}12\\ \times 7\\ \hline 84\end{array}$	$\begin{array}{r}12\\ \times 8\\ \hline 96\end{array}$	$\begin{array}{r}12\\ \times 9\\ \hline 108\end{array}$

Your Turn Problem #6
Perform the indicated division.

$6,384 \div 12$

| 12 | 12 | 12 |
| $\times\, 1$ | $\times 2$ | $\times\, 3$ |

| 12 | 12 | 12 |
| $\times\, 4$ | $\times\, 5$ | $\times\, 6$ |

| 12 | 12 | 12 |
| $\times\, 7$ | $\times\, 8$ | $\times\, 9$ |

Answer:_____

Division with Remainders

For this section, when a division problem does not work out evenly, write the remainder preceded by an "R". Later we will perform division with fractions and decimals. **You can check by multiplying the divisor and the quotient and adding the remainder.** This result should equal the dividend.

Example 7a: $245 \div 8$.   Solution:

$$
\begin{array}{r}
30 \\
8\overline{)245} \\
-24 \\
\hline
05 \\
-0 \\
\hline
5
\end{array}
$$

Answer: 30 R 5

Check

$$
\begin{array}{r}
30 \\
\times\, 8 \\
\hline
240
\end{array}
$$

$$
\begin{array}{r}
240 \\
+\quad 5 \\
\hline
245
\end{array}
$$ It checks.

Writing all of the products before dividing.

Example 7b: $4680 \div 17$	
1. First rewrite in long division format. $17\overline{)4680}$	
2. Write the products of the divisor, 17, and the numbers 1-9 (shown to the right).	$\begin{array}{r}17\\ \times1\\ \hline 17\end{array}$ $\begin{array}{r}17\\ \times2\\ \hline 34\end{array}$ $\begin{array}{r}17\\ \times3\\ \hline 51\end{array}$
3. Determine where the first number will be placed in the quotient (above the division box). 17 will not divide into 4 but it will divide into 46. Therefore the first digit of the quotient is written above the 6.	$\begin{array}{r}17\\ \times4\\ \hline 68\end{array}$ $\begin{array}{r}17\\ \times5\\ \hline 85\end{array}$ $\begin{array}{r}17\\ \times6\\ \hline 102\end{array}$
4. Looking at our products to the right, we want to find the product closest to 46 without going over. Answer: 2. Write the 2 in the quotient, multiply, then subtract and bring down the next digit. $\begin{array}{r}2\\ 17\overline{)4680}\\ -34\\ \hline 128\end{array}$	$\begin{array}{r}17\\ \times7\\ \hline 119\end{array}$ $\begin{array}{r}17\\ \times8\\ \hline 136\end{array}$ $\begin{array}{r}17\\ \times9\\ \hline 153\end{array}$
5. Do it again. Find the product to the right closest to 128 without going over. Answer: 7. $\begin{array}{r}27\\ 17\overline{)4680}\\ -34\\ \hline 128\\ -119\\ \hline 90\end{array}$	
6. Do it again. Find the product to the right closest to 90 without going over. Answer: 5. $\begin{array}{r}275\\ 17\overline{)4680}\\ -34\\ \hline 128\\ -119\\ \hline 90\\ -85\\ \hline 5\end{array}$ **Answer:** $\boxed{275\text{ R }5}$	

Your Turn Problem #7

Perform the indicated division.

$495 \div 7$

Answer: _____

Definition: Fraction (or a Rational Number)

A rational number is a number that can be written in the fraction form $\frac{a}{b}$ where 'a' is whole number and b is a whole number that is not zero. The top number is called the numerator and the bottom number is called the denominator.

$\frac{a}{b}$ $\leftarrow$ numerator
$\leftarrow$ denominator

Examples:: $\frac{3}{4}, \frac{5}{7}, \frac{87}{100}$

Note: In later courses, we will expand on the types of numbers a rational number can have in the numerator and denominator.

A fraction can be used to indicate division.

Since $\frac{a}{b}$ can be described as $a \div b$, b cannot be zero since division by zero is not possible--it is

undefined. $\left(\frac{a}{b} \rightarrow b\overline{)a} \quad \text{or} \quad \text{denominator}\overline{)\text{numerator}} \right)$

Example: $\frac{12}{3}$ is equivalent to $12 \div 3$ which equals 4.

Example 8a. Simplify: $\frac{28}{4}$

Answer: $\frac{28}{4}$ is equivalent to $28 \div 4$ which equals 7.

Example 8b. Simplify: $\dfrac{55}{11}$

Answer: $\dfrac{55}{11}$ is equivalent to $55 \div 11$ which equals 5.

Example 8c. Simplify: $\dfrac{48}{3}$

Answer: $\dfrac{48}{3}$ is equivalent to $48 \div 3$ which equals 16.

Your Turn Problem #8

Simplify the following.

a) $\dfrac{75}{5}$ b) $\dfrac{21}{3}$ c) $\dfrac{36}{2}$

Answer: _____   Answer: _____   Answer: _____

Recall: Any number divided by 1 is the same number. Any number divided by itself equals one.

Example 9a. Simplify: $\dfrac{13}{1}$

Answer: $\dfrac{13}{1}$ is equivalent to $13 \div 1$ which equals 13.

Example 9b. Simplify: $\dfrac{24}{24}$

Answer: $\dfrac{24}{24}$ is equivalent to $24 \div 24$ which equals 1.

Example 9c. Simplify: $\dfrac{12}{1}$

Answer: $\dfrac{12}{1}$ is equivalent to $12 \div 1$ which equals 12.

Your Turn Problem #9

Simplify the following.

a) $\dfrac{14}{14}$

b) $\dfrac{32}{1}$

c) $\dfrac{8}{8}$

Answer: _____

Answer: _____

Answer: _____

Recall: Any number divided by 0 is *undefined*. If zero is divided by any number except zero, it is equal to zero. If both dividend and divisor are zero, then the fraction is *indeterminate*.

Example 10a. Simplify: $\dfrac{8}{0}$

Answer: $\dfrac{8}{0}$ is equivalent to $8 \div 0$ which is undefined.

Example 10b. Simplify: $\dfrac{0}{4}$

Answer: $\dfrac{0}{4}$ is equivalent to $0 \div 4$ which equals 0.

Example 10c. Simplify: $\dfrac{0}{10}$

Answer: $\dfrac{0}{10}$ is equivalent to $0 \div 10$ which equals 0.

Your Turn Problem #10

Simplify the following if possible.

a) $\dfrac{15}{0}$

b) $\dfrac{19}{0}$

c) $\dfrac{0}{2}$

d) $\dfrac{0}{0}$

Answer: _____

Answer: _____

Answer: _____

Answer: _____

2.2 Homework: Division of Whole Numbers

Divide the following if possible.

1. $13 \div 0$

2. $0 \div 9$

3. $0 \div 0$

4. $15 \div 15$

5. $7 \div 7$

6. $141 \div 3$

7. $11 \div 0$

8. $0 \div 11$

9. $735 \div 5$

10. $1524 \div 6$

2.2 Homework: Division of Whole Numbers cont.

Divide the following if possible.

11. 2936 ÷ 8

12. 4887 ÷ 9

13. 93 ÷ 0

14. 0 ÷ 71

15. 1842 ÷ 6

16. 67,921 ÷ 7

17. 0 ÷ 0

18. 16 ÷ 0

19. 138 ÷ 6

20. 324 ÷ 6

2.2 Homework: Division of Whole Numbers cont.
Divide the following if possible.

21. $756 \div 9$

22. $2{,}695 \div 4$

23. $45{,}967 \div 3$

24. $5{,}648 \div 8$

25. $1{,}068 \div 53$

26. $51{,}975 \div 15$

27. $97{,}356 \div 42$

28. $414 \div 13$

2.2 Homework: Division of Whole Numbers cont.
Divide the following if possible.

29. $52,345 \div 24$

30. $129,212 \div 16$

31. $374 \div 0$

32. $0 \div 987$

33. $5,324 \div 78$

34. $192,211 \div 18$

2.2 Homework: Division of Whole Numbers cont.

Divide the following if possible.

35. $567 \div 11$

36. $792 \div 24$

37. $887 \div 18$

38. $5461 \div 43$

39. $264 \div 14$

40. $156 \div 30$

41. $0 \div 0$

42. $33 \div 0$

2.2 Homework: Division of Whole Numbers cont.

Solve the following applications.

43. There are 517 students who are taking a field trip. If each bus can hold 42 students, how many buses will be needed for the field trip?

44. Construction of a fence section requires 8 boards. If you have 260 boards available, how many sections can you build?

45. Mark drove 408 miles on 12 gallons of gas. Find the mileage (in miles per gallon) of Mark's car.

46. A machine can produce 240 bottles per minute. How long will it take to produce 25,680 bottles?

47. A large container has 735 liters of a liquid. How many 2-liter bottles can be filled from the 735 liters?

2.2 Homework: Division of Whole Numbers cont.

Simplify the following fractions if possible.

48. $\dfrac{48}{4}$

49. $\dfrac{120}{5}$

50. $\dfrac{27}{27}$

51. $\dfrac{395}{395}$

52. $\dfrac{0}{9}$

53. $\dfrac{0}{12}$

54. $\dfrac{80}{5}$

55. $\dfrac{93}{0}$

56. $\dfrac{0}{22}$

57. $\dfrac{0}{0}$

58. $\dfrac{11}{11}$

59. $\dfrac{56}{2}$

2.3 Exponents

Exponential Form

A form of writing the product of a factor that repeats

Example: $2 \cdot 2 \cdot 2 = 2^3$ or $2^3 = 2 \cdot 2 \cdot 2$

Base: The base is the factor being repeatedly multiplied

Exponent: The exponent is a small number written to the right of the base and raised a half space. It indicates the number of bases being multiplied.

Exponent (or power)

Base $\longrightarrow 2^3$

How to say it.

Exponent: 2 – squared or " to the second power"

3 – cubed or "to the 3rd power"

4 – to the fourth power

etc.

Examples:

2^3: two cubed or 2 to the 3rd (power)

8^2: eight squared or 8 to the 2nd (power)

3^7: three to the seventh (power)

The word "power" is often left out.

Procedure: To evaluate a number in exponential form

Step 1: Rewrite as repeated multiplication.

Step 2: Multiply the repeated multiplications.

Example 1. a) Evaluate: 3^4 b) Evaluate: 5^2

Solutions:
$$3^4 = 3 \cdot 3 \cdot 3 \cdot 3$$
$$= \boxed{81}$$

$$5^2 = 5 \cdot 5$$
$$= \boxed{25}$$

c) Evaluate: 7^3 d) Evaluate: 2^5

Solutions:
$$7^3 = 7 \cdot 7 \cdot 7$$
$$= \boxed{343}$$

$$2^5 = 2 \cdot 2 \cdot 2 \cdot 2 \cdot 2$$
$$= \boxed{32}$$

Your Turn Problem #1

a) Evaluate 5^3 _____

b) Evaluate 7^2 _____

c) Evaluate 3^3 _____

d) Evaluate 2^4 _____

Exponents of One

A number with an exponent of '1' is simply equal to the number.

Example 2a. Evaluate: 8^1

Answer: $8^1 = 8$

Example 2b. Evaluate: 135^1

Answer: $135^1 = 135$

Your Turn Problem #2

a) Evaluate 4^1 _____

b) Evaluate 76^1 _____

Exponents of Zero

A number with an exponent of "0" is equal to 1. It doesn't make too much sense right now. The explanation will come later once we have learned a few algebra tools.

Example 3a. Evaluate: 8^0 **Example 3b.** Evaluate: 95^0

Answer: $8^0 = 1$

Your Turn Problem #3

a) Evaluate 11^0 _____   b) Evaluate 25^0 _____

Operations involving two or more numbers in exponential form

Step 1: Evaluate each number in exponential form.

Step 2: Perform any other operations.

Example 4. Evaluate: $2^5 + 7^2$

Solution: First rewrite each as repeated multiplication $2^5 + 7^2 = 2 \cdot 2 \cdot 2 \cdot 2 \cdot 2 + 7 \cdot 7$

Multiply the repeated multiplications. $32 + 49$

Perform any other operations.   **Answer:** $\boxed{81}$

Your Turn Problem #4

Evaluate $4^3 + 5^3$

Answer: _____

Example 5. Evaluate: $3^2 \cdot 4^2 \cdot 1^5$

Solution: First rewrite each as repeated multiplication: $3^2 \cdot 4^2 \cdot 1^5 = 3 \cdot 3 \cdot 4 \cdot 4 \cdot 1 \cdot 1 \cdot 1 \cdot 1 \cdot 1$

Multiply the repeated multiplications. $9 \cdot 16 \cdot 1$

Perform any other operations. **Answer:** $\boxed{144}$

Note: If the base is 1, it doesn't matter what the exponent is. The result will always be 1.

Your Turn Problem #5

Evaluate: $5^3 \cdot 3^3 \cdot 1^9$

Answer: _____

Evaluating with a base of 10

Example 6a. Evaluate: 10^2

Solution: First rewrite as repeated multiplication, then multiply.

$$10^2 = 10 \cdot 10$$

Answer: $\boxed{100}$

Example 6b. Evaluate: 10^3

Solution: First rewrite as repeated multiplication, then multiply.

$$10^3 = \underbrace{10 \cdot 10}_{100} \cdot 10$$

$$100 \cdot 10$$

Answer: $\boxed{1000}$

Note that the pattern. The number of 10's being multiplied is the same as the number of zeros in answer. Similarly, the number of zeros in the answer is equal to the exponent.

$$10^1 = 10 \text{ (one zero, same as exponent)}$$

$$10^2 = 100 \text{ (two zeros, same as exponent}$$

$$10^3 = 1000 \text{ (three zeros, same as exponent)}$$

Using this pattern for 10^8; the answer would be the number "1" followed by eight "0"s.

$$10^8 = 100000000$$

Then starting from the right, place commas after every third digit.

$$100,000,000$$

Your Turn Problem #6

Evaluate: 10^5

Answer: _____

Procedure: Evaluating with a base of 10

Step 1: Write the number one, and then insert an amount of zeros equal to the exponent.

Step 2: Perform any other operations and then starting from the right, place commas after every third digit.

Example 7. Evaluate: $7 \cdot 10^5$

Solution: $7 \cdot 10^5$

                 $7 \cdot 100,000$ Step 1

Answer: $\boxed{700,000}$ Step 2

Your Turn Problem #7

Evaluate: $9 \cdot 10^3$

Answer: _____

Example 8. Evaluate: $2^2 \cdot 3^2 \cdot 10^3$

Solution: $\qquad\qquad 2^2 \cdot 3^2 \cdot 10^3$

$\qquad\qquad\qquad 2 \cdot 2 \cdot 3 \cdot 3 \cdot 1000 \qquad\qquad$ Step 1

$\qquad\qquad\qquad 4 \cdot 9 \cdot 1000$

Answer: $\qquad \boxed{36,000} \qquad\qquad$ Step 2

Your Turn Problem #8

Evaluate: $5^2 \cdot 2^3 \cdot 10^4$

Answer: _____

2.3 Homework: Exponents

Evaluate the following

1. 2^4

2. 3^3

3. 10^2

4. 5^3

5. 15^0

6. 23^1

7. 7^3

8. 12^2

9. 0^5

10. 33^0

11. 6^3

12. 3^5

2.3 Homework: Exponents cont,

Evaluate the following.

13. 11^0

14. 9^1

15. 1^{21}

16. 6^0

17. 8^2

18. 10^4

19. 7^2

20. 2^7

21. 1^{18}

22. 2^0

23. 4^3

24. 14^0

2.3 Homework: Exponents cont,

Evaluate the following.

25. $8^2 + 3^3$

26. $2^5 - 5^2$

27. $3^4 - 3^3$

28. $5^0 + 8^0$

29. $19^0 - 15^0$

30. $5^1 \cdot 3^0$

31. $2^2 \cdot 3^2 \cdot 4^2$

32. $1^{15} + 1^{13}$

33. $8^0 \cdot 2^4$

34. $7^0 \cdot 4^3$

35. $1^8 \cdot 3^3 \cdot 2^4$

36. $1^5 \cdot 2^3 \cdot 9^0$

2.3 Homework: Exponents cont,

Evaluate the following.

37. $12 \cdot 10^4$

38. $2^5 \cdot 10^3$

39. $3^4 \cdot 2^2 \cdot 10^5$

40. $0^2 \cdot 3^2 \cdot 10^2$

41. $9^0 + 11^1$

42. $10^2 + 18^0$

43. $1^{35} + 23^0$

44. $19^0 - 35^0$

2.4 Order of Operations

Order of Operations Agreement

Given a problem such as $5 \cdot 2^3 + 8 \div 2$, there are several different answers that can be found if we do not have any rules to follow. Fortunately, we have those rules. To keep it all straight, it is important that you memorize the following-- the Order of Operations Agreement. Please note the first letter of each step. The Order of Operations is a priority list. The operations higher up on the priority list get taken care of first.

Procedure: For Order of Operation

Step 1: Parentheses: perform operations inside parentheses or <u>other grouping symbols</u> (brackets or braces).

Step 2: Exponents: simplify (evaluate) exponential notation expressions

Step 3: Multiply or divide as they appear from left to right

Step 4: Add or subtract as they appear from left to right

When you are asked to "simplify" remember to use the Order of Operations Agreement (PEMDAS). In simplifying, work vertical; *for each step recopy the entire problem.* For each line of the problem, perform only the step being performed in the Order of Operations.

Restating the Rules:

1. Always simplify what is inside the parentheses or grouping symbols first.
2. If there is a number with an exponent, then evaluate the number with the exponent.
3. Multiplication and Division is performed before Addition and Subtraction.
4. If there is Multiplication or Division next to each other, do whichever is written first (left to right).
5. If there is Addition or Subtraction next to each other, do whichever is written first (left to right).

Example 1a. Simplify $13 + 2 \cdot 3$

$$13 + 6$$

Answer: $\boxed{19}$

Step 1. Parentheses: None

Step 2 Exponents: None

Step 3. Multiplication or Division

Step 4. Addition & Subtraction

Example 1b. Simplify $4 + 3(12 - 7)$

$$4 + 3(5)$$

$$4 + 15$$

Answer: $\boxed{19}$

Step 1. Parentheses

Step 2. Exponents: None

Step 3. Multiplication or Division

Step 4. Addition & Subtraction

Note: 3(5) in the example above indicates multiplication.
3[5] also indicates multiplication.

Example 1c. Simplify $12 - 5 + 2$

$$7 + 2$$

Answer: $\boxed{9}$

Step 1. Addition and Subtraction

(Remember left to right)

Your Turn Problem #1

a) Simplify $8 + 12 \div 3$ b) Simplify $5 + 3(7 - 2)$ c) Simplify $24 \div 2 \cdot 3$

Answer:_____   Answer:_____   Answer:_____

Essential Mathematics Workbook

Example 2a. Simplify $2^3 - 5 + 3^2$ Step 1. Parentheses: None

$$8 - 5 + 9$$ Step 2. Exponents

$$3 + 9$$ Step 3. Multiplication or Division: None

Answer: $\boxed{12}$ Step 4. Addition & Subtraction: left to right

Example 2b. Simplify $(4 + 2)^2 \div 2 \cdot 3$ Step 1. Parentheses

$$(6)^2 \div 2 \cdot 3$$ Step 2. Exponents

$$36 \div 2 \cdot 3$$ Step 3. Multiplication or Division

$$18 \cdot 3$$ (Left to Right)

Answer: $\boxed{54}$

Your Turn Problem #2

a) Simplify $(5 + 2) + 3(2^3 - 5)$ b) Simplify $(2 + 1)^2 - (12 \div 4 \cdot 3)$

Answer:_____   Answer:_____

Example 3a. Simplify $2 + [8 + 3(5 - 2)]$

$$2 + [8 + 3(3)]$$

$$2 + [8 + 9]$$

$$2 + [17]$$

Answer: $\boxed{19}$

Step 1. Parentheses: first, do inside the inner grouping symbols, parentheses.

Then simplify inside the brackets.

Step 2. Addition

Example 3b. Simplify $2 + 5[12 - 4(5 - 3)]$

$$2 + 5[12 - 4(2)]$$

$$2 + 5[12 - 8]$$

$$2 + 5[4]$$

$$2 + 20$$

Answer: $\boxed{22}$

Step 1. Parentheses: first inside parentheses.

Then simplify inside brackets.

Step 2. Multiplication

Step 3. Addition

Your Turn Problem #3

a) Simplify $3 + 4[12 - (11 - 5)]$

b) Simplify $20 - 2[12 \div (6 - 2)]$

Answer:_____

Answer:_____

2.4 Homework: Order of Operations

Simplify the following.

1. $12-3+5$

2. $12 \div 3 \cdot 4$

3. $5+3 \times 2$

4. $18-12 \div 3$

5. $18 \div 3 \times 2$

6. $24 \div 6 \times 2$

7. $(12+3) \div 5$

8. $18-3 \times 2+4$

9. $2^3 \div 2$

10. $(7-4)^2$

2.4 Homework: Order of Operations cont.

Simplify the following.

11. $(15 + 2 - 8)^2$

12. $(4 + 2 - 1)^3 \div 12^0$

13. $(3 + 1)^4 \div 2$

14. $5^2 - 16 \div 2 + 8$

15. $(5 - 3)^3 \times 12$

16. $(10 - 7)^2 + (11 - 9)^3$

17. $(10 - 8 + 1)^4$

18. $(8 \div 4 \times 2)^3$

19. $48 \div (5 - 4 + 7)$

20. $36 \div (2^3 - 7 + 1)$

2.4 Homework: Order of Operations cont.

Simplify the following.

21. $(14-4)^0$

22. $36 \div \left(2^4 - 3^2 + 5^1\right)$

23. $30 \div 6 + 12 \div 2$

24. $36 \div 6 \times 2 + \left(2^4 - 5^0\right) \div 3$

25. $(15 \div 5 \times 3) \div (7-6)^{21}$

26. $(5-3)^4 - (9 \div 3)^2 + (4 \times 7)^0$

27. $3 + 2(5)$

28. $2^3 + 4(13-5)$

29. $5[13 - (10-7)]$

30. $2[20 - (15-9)]$

2.4 Homework: Order of Operations cont.

Simplify the following.

31. $20 + [12 - (8 - 5)]$

32. $3 + 2[18 - (12 - 8)]$

33. $8 - 3[7 - (3 + 2)]$

34. $5 + 2[(12 + 8) \div 4]$

35. $5 \times [(4 + 2)^2 \div 6 \times 4]$

36. $7 \times [5^2 - (3 + 1)^2]$

37. $(7 + 3)[5 + 2(3 + 2)]$

38. $(3 + 4)^2 + 8 \cdot 3$

39. $8 + 2[(4 + 2)^2 \div 3]$

40. $2 + 3[24 \div (8 - 5 + 3)]$

Practice Test 2

Name the property that is being illustrated. (Do not evaluate. Just name the property)

1. $7 \cdot 3 = 3 \cdot 7$:_____

2. $12 \cdot 1 = 12$:_____

3. $2 \cdot 0 = 0$:_____

4. $6 \cdot (3 \cdot 4) = (6 \cdot 3) \cdot 4$:_____

5. $7 \cdot (9 + 3) = (7 \cdot 9) + (7 \cdot 3)$:_____

Perform the indicated operation.

6. 27×9

7. 48×37

8. 405×79

9. Find the product of 67 and 39.

Perform the indicated division if possible.

10. $9 \div 0$

11. $0 \div 7$

12. $18 \div 18$

13. $13 \div 1$

Practice Test 2 cont.

Perform the indicated division if possible.

14. $1,968 \div 8$

15. $45,216 \div 9$

16. $8,276 \div 23$

17. $237 \div 7$

18. $2,514 \div 12$

19. $9,073 \div 35$

Evaluate the following expressions.

20. 3^4

21. 7^0

22. 9^1

23. 1^{27}

24. $(11 - 9)^3$

25. $(3 + 1)^2 \times 3$

Practice Test 2 cont.

Evaluate the following expressions.

26. $5^3 + \left(3^2 - 2^3\right)$

27. $\left(12 - 8 + 2\right)^2 \times 5$

28. $2^3 + 24 \div 3$

29. $24 \div 4 \times 6$

30. $20 - \left(8 \div 2\right)^2$

31. $40 - \left(6 \div 2 \times 3\right) + 5$

32. $\left(2^4 + 2^3\right) \div 3$

33. $24 \div 12 + 18 \div 3$

34. $20^1 \div 1^{20}$

35. $\left(24 \div 12 \times 2\right)^3 + 3^2$

3.1 Solving Equations

Equation

A sentence with an "=" (equal sign) is called an equation.

Examples: $9 + 3 = 12$; $15 - 8 = 7$. $x + 5 = 19$, $3 \cdot N = 51$

An equation is a sentence.

Using the example $9 + 3 = 12$; $9 + 3 = 12$ translates to "9 and 3 is 12". The word "and" is a conjunction. 9 and 3 are the subjects. "Is" is the verb. 12 is the predicate nominative.

Solution

An equation may be written with one or more of the numbers not known.

A **solution** of an equation is a number that would make the sentence true if one of the numbers is not known.

Example: 7 plus what number is 16?

Using symbols: $7 + \boxed{} = 16$

The solution to this equation would be 9 because $7 + 9 = 16$.

Variable

We usually use a letter instead of a blank box. The letter is called a variable.

Example: $7 + N = 16$

Solutions of an Equation

The solution of an equation is a number which when replaced for the variable results in a true sentence. The left side of the equal sign is the same as the right side.

Solving Equations

We will begin to solve certain types of equations that are pertinent to the topics of this course.

There may be occasions where mentally you can see what the answer is. That's good. However, try to follow a process if given. When the equations get more complicated, you will have practiced the process and the solution will be easier to find.

Equations where the Variable is Isolated

The first equation to consider is an equation where the variable is by itself on one side of the equal sign. The procedure is to perform the operation(s) on the other side of the equal sign.

Example 1a. Solve: $x = 342 \div 3$

Solution: Perform the division on the right side of the "=" sign.

$$
\begin{array}{r}
114 \\
3\overline{)342} \\
-3 \\
\hline
04 \\
-3 \\
\hline
12 \\
-12 \\
\hline
0
\end{array}
$$

Answer: $\boxed{x = 114}$

Example 1b. Solve: $x = 500 - 192$

Solution: Perform the division on the right side of the "=" sign.

$$
\begin{array}{r}
4\ 9 \\
5\ 0^1 0 \\
-1\ 9\ 2 \\
\hline
3\ 0\ 8
\end{array}
$$

Answer: $\boxed{x = 308}$

Your Turn Problem #1

Solve the following equations.

a) $N = 30 \cdot 500$

b) $N = 8460 \div 12$

Answer:_____

Answer:_____

<u>Solving Equations of the form x + a = b</u>

Let's look at an example in this form.

$x + 5 = 8$

What number plus 5 equals 8. The answer is 3.

Many students can often visualize what this answer is without showing any work, but we will take advantage of knowing what the answer is and form a procedure.

An equation is a balance. If you add a number to one side, then you must add the same number to the other side to keep the balance. Likewise, if you subtract a number on one side, you must subtract the same number on the other side. Therefore, if we subtract 5 on both sides, we will end up with just x on the left hand side and our 3 on the right hand side. Keep in mind, learning the proper procedures will make your transition to algebra courses much smoother.

Procedure: To solve an equation of the form x + a = b

1. Subtract 'a' on both sides
2. Simplify left hand side and the right hand side.

Example 2a. Solve: $x + 18 = 24$

Solution:	$x + 18 - 18 = 24 - 18$	Subtract 18 on both sides.
	$x + 0 = 6$	$(18 - 18 = 0 \text{ and } 24 - 18 = 6)$
Answer:	$\boxed{x = 6}$	(on the left hand side of the "=",
		$x + 0 = x)$

Example 2b. Solve: $x + 95 = 200$

Solution: $\qquad x + 95 - 95 = 200 - 95 \qquad\qquad$ Subtract 95 on both sides.

$\qquad\qquad\qquad x + 0 = 105 \qquad\qquad\qquad$ $(95 - 95 = 0$ and $200 - 95 = 105)$

$$\begin{array}{r} 1\ 9 \\ 2\ \cancel{0}^{1}0 \\ -\ \ 9\ 5 \\ \hline 1\ 0\ 5 \end{array}$$

Answer: $\qquad\boxed{x = 105}\qquad\qquad\qquad\qquad (x + 0 = x)$

Note: Subtracting the number on the same line is a horizontal method. This process could also have been performed vertically. Choose the method you feel most comfortable with.

Example 2c. Solve: $x + 12 = 28$

Solution: $\qquad \begin{array}{r} x + 12 = \ \ 28 \\ -12 \quad -12 \\ \hline x + 0 = \ \ 16 \end{array} \qquad$ Subtract 12 on both sides.
$\qquad\qquad\qquad\qquad\qquad\qquad (12 - 12 = 0$ and $28 - 12 = 16)$

Answer: $\qquad\boxed{x = 16}$

Example 2d. Solve: $9 + N = 30$

Solution: $\qquad 9 + N - 9 = \ 30 - 9 \qquad$ Subtract 9 on both sides.

$\qquad\qquad\quad N + 9 - 9 = \ 21 \qquad\qquad (9 + N$ is the same as $N + 9$; Comm. Prop.)

$\qquad\qquad\qquad\quad N + 0 = \ 21 \qquad\qquad\qquad 12 - 12 = 0$

Answer: $\qquad\boxed{N = 21}$

Your Turn Problem #2

Solve the following equations.

a) $x + 134 = 700$ $\qquad\qquad$ b) $x + 46 = 93$ $\qquad\qquad$ c) $28 + A = 54$

a) Answer:_____ $\qquad$ b) Answer:_____ $\qquad$ c) Answer:_____

Solving Equations of the form a · x = b

Let's look at an example in this form.

$$3 \cdot x = 24$$

3 times what number equals 24? The answer is 8.

Remember, an equation is a balance. If you multiply a number to one side, then you must multiply the same number to the other side to keep the balance. Likewise, if you divide a number on one side, you must divide the same number on the other side. Therefore, if we divide 3 on both sides, we will end up with just x on the left hand side and our 8 on the right hand side.

We are going to use a fraction notation to indicate division. We used this notation previously in the division section. Let's review.

Example 3a. Simplify: $\dfrac{36}{4}$

Answer: $\dfrac{36}{4}$ is equivalent to $36 \div 4$ which equals 9.

Example 3b. Simplify: $\dfrac{6252}{12}$

Answer: $\dfrac{6252}{12}$ is equivalent to $6252 \div 12$ which equals 521.

Example 3c. Simplify: $\dfrac{9}{9}$

Answer: $\dfrac{9}{9}$ is equivalent to $9 \div 9$ which equals 1. (Any number divided by itself equals 1.)

Your Turn Problem #3

Simplify the following.

a) $\dfrac{72}{4}$

b) $\dfrac{15}{15}$

c) $\dfrac{0}{8}$

Answer: _____

Answer: _____

Answer: _____

Procedure: To solve an equation of the form a · x = b

1. Divide by 'a' on both sides. We will use fraction notation to indicate division.

    We will show this step by drawing a line under each side separately for division.

    Then write the number next to the variable under each line.

2. Simplify left hand side and the right hand side.

Note: The answer will always be a whole number without a remainder. Later, we will cover equations where the answer is not a whole number.

Example 4a. Solve: $3 \cdot x = 24$

Solution: $3 \cdot x = 24$

$$\frac{3 \cdot x}{3} = \frac{24}{3}$$ Divide by 3 on both sides.

$$\frac{\cancel{3} \cdot x}{\cancel{3}} = \frac{24}{3}$$ Simplify each side. Any number divided by itself equals 1.

Answer: $\boxed{x = 8}$ $(1 \cdot x = x)$

Example 4b. Solve: $7 \cdot x = 238$

Solution:

$$7 \cdot x = 238$$

$$\frac{7 \cdot x}{7} = \frac{238}{7} \qquad \text{Divide by 7 on both sides.}$$

$$\frac{\mathbf{7} \cdot x}{\mathbf{7}} = \frac{238}{7} \qquad \text{Simplify each side. Any number divided by itself equals 1.}$$

Answer: $\boxed{x = 34}$

Your Turn Problem #4

Solve the following equations.

a) $9 \cdot x = 432$ 　　　　　　　　 b) $15 \cdot x = 3045$

a) Answer:_____ 　　　　　　 b) Answer:_____

Equations written "backwards"

Some equations may be written with the variable on the right hand side of the "=" sign.

Example: Solve $15 = x + 3$

Our goal is still the same, we need to get 'x' by itself. So always subtract the number next to x on each side.

$$15 = x + 3$$
$$15 - 3 = x + 3 - 3$$
$$12 = x \quad \text{This can be written as } x = 12.$$

Rewriting an equation so that the x is on the left hand side is a valid step given by the symmetric property.

Symmetric Property: If $b = a$, then $a = b$

The symmetric property lets us rewrite the equation with the 'x' on the left hand side. This is an optional step. Some students find it easier to have x on the left hand side as soon as possible. If you wish to solve the equation without using the symmetric property, it will still work out fine.

Example 5a. Solve: $45 = x + 7$

Solution:

$$x + 7 = 45 \qquad \text{Symmetric Property}$$
$$\underline{-7 \quad -7} \qquad \text{Subtract 7 on both sides.}$$
$$x + 0 = 38$$

Answer: $\boxed{x = 38}$

Example 5b. Solve: $225 = 3 \cdot x$

Solution:

$$3 \cdot x = 225 \qquad \text{Symmetric Property}$$

$$\frac{3 \cdot x}{3} = \frac{225}{3} \qquad \text{Divide by 3 on both sides.}$$

$$\frac{3 \cdot x}{3} = \frac{225}{3} \qquad \text{Simplify each side. Any number divided by itself equals 1.}$$

Answer: $\boxed{x = 75}$

Your Turn Problem #5

Solve the following equations.

a) $71 = x + 39$ 　　　　　　　　b) $2665 = 13 \cdot x$

a) Answer:_____ 　　　　b) Answer:_____

3.1 Homework: Solving Equations

Solve the following equations.

1. $x + 5 = 19$

2. $x + 86 = 100$

3. $N = 51 - 48$

4. $N = 72 \div 2$

5. $5 \cdot x = 135$

6. $8 \cdot a = 176$

7. $x + 13 = 200$

8. $x + 77 = 88$

9. $N = 24 \times 19$

10. $Y = 300 - 124$

11. $7 \cdot x = 273$

12. $12 \cdot c = 600$

3.1 Homework: Solving Equations cont.

Solve the following equations.

13. $x + 27 = 90$

14. $x + 345 = 2000$

15. $A = 30 - 12 - 5$

16. $B = 67 - 34 + 15$

17. $17 \cdot x = 1428$

18. $14 \cdot b = 14$

19. $65 + x = 75$

20. $43 + x = 115$

21. $22 = C + 5$

22. $148 = x + 52$

3.1 Homework: Solving Equations cont.

Solve the following equations

23. $391 = 17 \cdot x$

24. $896 = 16 \cdot b$

25. $380 \times 200 = N$

26. $75 - 25 = H$

27. $x = 20 - (5 + 8)$

28. $x = 35 - 15 + 5$

29. $x = 5^2 + 3 \times 4$

30. $x = (97 + 74 + 87) \div 3$

3.2 Word Problems using Whole Numbers

Students often ask the question "Why do I need to learn all this math?" The answer is simple. Math was created to solve applications (word problems). The applications will only consist of material that you have learned. Thus far, the material has consisted of operations with whole numbers. Therefore, the applications will consist of operations with whole numbers. Once we cover fractions, the applications will use your knowledge of operations with fractions. Applications will be covered in all of your math classes. Success in solving word problems will be obtained by following a process for solving word problems. Certainly, many students will say "I can do this in my head without showing any steps." That is great. However, we are also preparing for word problems that you will not be able to do without a procedure to follow.

Reading

Solving word problems begins with reading. We need to comprehend what information is being given and what is being asked. Take your time when you read the problem. Maybe reread it again to make sure you have read it correctly. Write down information while you are reading the problem. If there is a formula that applies, write it down as well. Identify keywords used to indicate the operation.

Let Statement

Once you know what is being asked, write down a "let statement." Assign a letter to the question being asked. Any letter may be used. We tend to use letters like x, y and n. For example: How many hours will it take to drive to Monterey? The letter (variable) assigned could be H for hours.

H = number of hours it will take to drive to Monterey.

Equation

Translate the problem into an equation. Use the variable from the let statement. This step may seem unnecessary at times, however, we are looking at the big picture here. We want to learn the correct process now. Not later.

Solve and Answer

Solve the equation. Then write your answer in a sentence. Make sure the answer "makes sense." If the question is asking for the price of a textbook, and your answer is $8000, it doesn't make sense. If your answer is $95, that makes sense. It's still expensive, but it makes sense.

General Steps to Solving Word Problems

1. *Read, Write, and Identify.* Read and reread the problem carefully making note of all data (numbers) and keywords. Write down the information. Make a chart, picture, or diagram if possible.

2. *Let Statement.* Identify the question? What is being asked? Choose a symbol for this unknown.

3. *Equation.* Identify the operation to be used (addition, subtraction, multiplication, or division). Translate the problem into an equation using the variable.

4. *Solve.* Solve the equation.

5. *Answer.* Answer the question in a sentence. Make sure proper units ($, feet, hours, books, etc.) are used and the answer makes sense.

Key phrases

Memorizing the following key phrases will assist you in determining the operation to be performed in a word problem.

Addition

Sum , total, increased by, more than, added to, plus

Recall: Addition is commutative. The order does not matter. $5 + 12 = 12 + 5$

Subtraction

Minus, Subtract, Take away, Difference , Subtracted from *, Less than *,

Decreased by, How Much More, Taken from, Remains, Deduct, Reduced by, Leftover

Recall: Subtraction is not commutative. $12 - 5$ is not the same as $5 - 12$.

* For the phrases, "subtracted from" and "less than", the order must be reversed.

13 subtracted from 20 translates to $20 - 13$.

Likewise, x subtracted from y translates to $y - x$. (The second letter or number is written first.)

Multiplication

Product, Times, Multiplied by, Twice (2 × __), By (10 ft by 12 ft room)

Multiplication is commutative. The order does not matter. $5 \times 12 = 12 \times 5$

Multiplication is also associative: $5 \times 7 \times 3 = 105$(doesn't matter which you multiply first.)
$$35 \times 3 = 105 \text{ or } 5 \times 21 = 105$$

Division

Quotient, Divided by/into, Ratio, Per, Split Evenly, how many ____ in ____?

(A total is separated into equal groups)

Example 1. The yearly profit for DR Construction was $78,216 in 2004, $153,917 in 2005, and $85,098 in 2006. What is the total profit for these three years?

Solution:
Step 1. The key word in this problem is "total".

Chart of yearly profit

year	2004	2005	2006	Total
profit	$78,216	$153,917	$85,098	?

Step 2. Let P = total profit for the three years.

Step 3. Equation: $P = \$78,216 + \$153,917 + \$85,098$

Step 4. Solve:

```
      7 8 2 1 6
    1 5 3 9 1 7
  +   8 5 0 9 8
  ─────────────
P = 3 1 7 2 3 1
```

Step 5. Answer question.

Answer: The total profit for the three years was $317,231.

Your Turn Problem #1

Laura's yearly salary is $58,500. Next year, her salary will increase by $7,000. What will her yearly salary be next year?

Data: _____

Key word:_____

Let S = _____

Equation:_____

Answer:_____

Example 2. In June, the Big Bear Boutique sold $24,760 worth of merchandise, but in July, it sold only $19,458 worth of merchandise. How much more did the boutique sell in June than in July?

Solution: "How much more" indicates subtraction.

Step 1. The key phrase in this problem is "how much more".

Chart of monthly sales

Month	June	July	Difference
Sales	$24,760	$19,458	?

Step 2. Let m = how much more was sold in June than in July.

Step 3. Equation: m = $24,760 − $19,458

Step 4. Solve:

$$\begin{array}{r} 24760 \\ -\ 19458 \\ \hline m=\ 5302 \end{array}$$

Step 5. Answer the question.

Answer: The boutique sold $5,302 more in June than July.

Your Turn Problem #2

The attendance for a concert was 12,329 on Friday and 23,421 on Saturday. How many more people attended on Saturday than on Friday?

Data: _____

Key word:_____

Let m =_____

Equation:_____

Answer:_____

A tip for the operation of <u>multiplication</u> is when a figure is given <u>per</u> some time period (sec., min., hr., day, wk., mo., yr.), and then the problem gives the number of the time period -- for example, the number of hours.

Example 3. Marcos agreed to buy a car paying $400 per month for 6 years. How much will he have paid after the 6 years?

Solution: keyword: "per" $400 per month for 6 years. 12 months per year.

6 years = ? months. 6×12 mo. = 72 months

$$\underbrace{400+400+400+...+400}_{72\text{months}}$$

Let A = amount Marcos will have paid after 6 years.

$A = 400 \cdot 72$

$A = 28{,}800$

Answer: Marcos will have paid $28,800 after 6 years.

Your Turn Problem #3

A Honda holds 17 gallons of gas. If the Honda gets 28 miles per gallon, how far can the Honda go on a full tank of gas?

Data: _____

Key word:_____

Let F = _____

Equation:_____

Answer:_____

Division defines the operation of finding how many groups of a certain number (the divisor) are contained in another number or amount (the dividend).

Example 4. A person earns $34,500 a year after taxes. How much does this person take home each month?

Solution: The total salary, $34,500 is split evenly into 12 equal groups. Therefore, this is a division problem.

?	?	?									

12 months

T = take home pay each month.

$T = 34,500 \div 12$

$T = \$2,875$

Answer: This person takes home each month $2,875.

Your Turn Problem #4

The receipts from a basketball game totaled $5,112. If each ticket was $9, then how many people attended the game?

Data: _____

Keyword:_____

Let ___ = _____

Equation:_____

Answer:_____

Average

The average of a set of numbers is the sum of the set numbers, divided by the number of addends.

Formula for Average:	$\dfrac{\text{Sum of a set of numbers}}{\text{number of addends}}$

Example 5. A student received scores of 72, 87, and 54. Find the average of the 3 test scores.

Solution: Average Problem, use formula.

Let A = average of the three test scores.

(Add the scores. Then divide by the number of scores, 3.)

$$A = \frac{72 + 87 + 54}{3}$$

$$A = \frac{213}{3} \quad \left(\text{Same as } 213 \div 3\right)$$

$$A = 71$$

Answer: The student has an average of 71 .

Your Turn Problem #5

Lori has 4 math tests with scores of 76, 85, 92, and 91. Find the average of all four tests.

Answer:_____

Example 6. On a four-day trip, Valentine drove the following number of miles each day: 240 miles, 360 miles, 118 miles, and 222 miles. What was the average number of miles she drove per day?

Solution: Let A = average of the number of miles driven per day.

(Add the number of miles driven per day. Then divide by the number of days, 4.)

$$A = \frac{240+360+118+222}{4}$$

$$A = \frac{940}{4} \quad \left(\text{Same as } 940 \div 4\right)$$

$$A = 235$$

Answer: Valentine drove an average of 235 miles per day.

Your Turn Problem #6

The temperature for a five - day period were as follows: 105°, 101°, 96°, 102°, 106°.

What was the average temperature for the five - day period?

Answer:_____

Example 7. Raul will purchase the following items at Home Depot: Three outdoor motion lights for $24 each, one patio table for $125, and 4 chairs for $39 each. If he only has $350 will he have enough money to purchase the items? (assume tax is already included) If there is enough money, how much money will he still have after the purchase? If he does not have enough money, how much more will he need?

Solution: Write down the data given: 3 lights @ $24

1 table @ $125

4 chairs @ $39 each

What is the total cost of the 8 items?

Let C = The total cost.

C = 3 × $24 + 1 × $125 + 4 × $39

C = $72 + $125 + $156

C = $353

Does Raul have enough money? He only has $350.

350 − 353 is not possible.

Answer: | No, Raul does not have enough money. He needs $3 to purchase all of the items. |.

Your Turn Problem #7

Daniel will purchase two packs of AA batteries for $7 each, three boxes of cereal for $4 each, and five boxes of pop tarts for $2 each. If he only as $40, will he have enough money to purchase these items? (Assume tax is already included) If there is enough money, how much money will he get back in change? If there is not enough how much more will he need?

Answer:_____

Example 8. Antonio's salary in 2010 was $42,000. If this is $5000 more than his salary in 2009, what was his salary in 2009?

Solution: Write down the data given:

2010 salary: $42,000

2009 salary is $5000 less than 2010 salary

Let s = Antonio's salary in 2009.

$s = 42000 - 5000$

$s = 37,000$

Answer: Antonio's salary in 2009 was $37,000.

Your Turn Problem #8

During a weekend series between the Dodgers and the Giants, where the Dodgers swept the Giants, the Dodgers scored 26 runs. If the Dodgers scored 8 more runs than did the Giants, how many runs did the Giants score?

Answer:_____

3.2 Homework: Word Problems using Whole Numbers

Solve the following applications.

1. Cathy spent $364 for tuition, $583 for books, and $35 for parking during one semester. What was the total cost for tuition, books, and parking for that semester?

2. Jacky's monthly paycheck of $879 was decreased by $175 for tax withholdings. What amount of pay did she receive after taxes?

3. Brian's checking account has a balance of $575. Brian wrote three checks for $54, $37, and $143. What was the new balance in the checking account?

4. A room contains 47 rows of seats. Each row has 28 seats. How many seats are in the room?

3.2 Homework: Word Problems using Whole Numbers cont.

Solve the following applications

5. Sandra drove 559 miles in her car using 13 gallons of gas. How many miles per gallon (mpg) does the car get?

6. Tim borrows $4,140. If he arranges to pay off the loan in 12 monthly payments, what is the monthly payment?

7. A computer printer can print 40 mailing labels per minute. How many labels can be printed in one hour?

8. Emile agrees to buy a car paying and $260 a month for 5 years. What is the total cost?
    (Hint: 5 years = 60 months)

3.2 Homework: Word Problems using Whole Numbers cont.

Solve the following applications

9. Kathie agrees to buy a car paying $300 down and $225 a month for 5 years. What is the total cost of the car?

10. There are 568 students who are taking a field trip. If each bus can hold 42 students, how many buses will be needed for the field trip?

11. Construction of a section of fence requires 24 boards. If you have 2600 boards available, how many full sections can you build?

12. Brandi earns $647 per week. How much money would she earn in 50 weeks?

3.2 Homework: Word Problems using Whole Numbers cont.

Solve the following applications

13. A student had test scores of 85, 72, 63, 91 and 94. Find the average of the test scores.

14 An employee's pay rate is $12 per hour. If she works 32 hours a week, how much will she make in two weeks?

15. Julie's records showed the following test scores in math: 96, 62, 72, and 82. What was her test score average?

16. Juan agrees to buy a car paying $1000 down and $456 a month for 6 years. What is the total cost of the car?

3.2 Homework: Word Problems using Whole Numbers cont.

Solve the following applications

17. A bottle company has 500 bottles of a certain type of liquid. If the machine separates the bottles into 8 bottle cases, how many cases can be created?

18. Sarah bought a laptop for $619, a laser printer for $194, and a desk for $144. If she only has $1000, will she have enough money to purchase these items? (Assume tax is already included) If there is enough money, how much money will she get back in change? If there is not enough how much more will she need?

19. There are 27 bones in a human hand and 26 bones in a human foot. How many bones are there in all in the hands and feet?

3.3 Geometric Applications - Perimeter

Perimeter of a Rectangle

The perimeter of a rectangle is the sum of all if its sides.

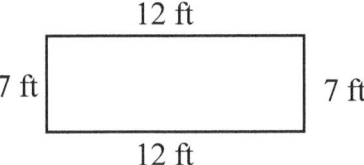

If we add all of the sides from the rectangle above, 7 ft + 7 ft + 12 ft + 12 ft = 38 ft.

Therefore, the perimeter of the rectangle above is 38 feet.

Typically, the length of the shorter side is called the width and the length of the longer side is called the length.

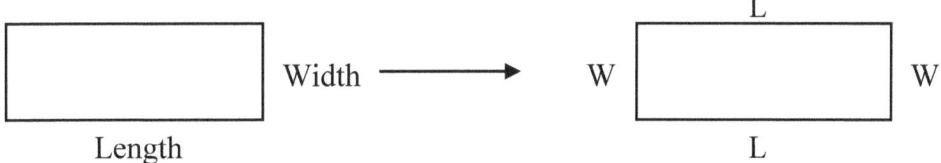

We can label the sides using L and W for length and width.

Formulas in math are often used describe a process. For instance to find the perimeter, we add all of the sides.

L + L + W + W = Perimeter.

L + L is the same as $2 \cdot L$ and W + W is the same as $2 \cdot W$.

So the formula for Perimeter of a Rectangle can be written as:

$2 \cdot L + 2 \cdot W = P$ or $P = 2 \cdot L + 2 \cdot W$

We can also omit the times symbol, and write $P = 2L + 2W$

If all these symbols got a little confusing, don't stress over it. For now, just remember the perimeter means to take the sum of all of its sides.

Perimeter of a Polygon

A polygon is a closed geometric figure with three or more sides. The perimeter of a polygon is the distance around it, or the sum of the lengths of its sides.

Please Note: The sketches in this section are not necessarily to scale.

Example: Find the perimeter of the polygon:

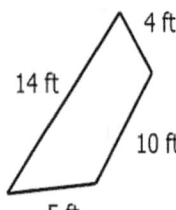

Solution: Add the lengths of all sides.

Perimeter = 5 ft + 10 ft + 4 ft + 14 ft

Answer: 33 ft

Finding a missing side.

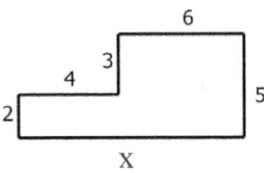

In this polygon to the left, there is only one side that is missing.
We will label x to show it is an unknown number.

Looking at the horizontal line segments, 4 and 6, we can determine what the bottom horizontal line segment is.

The missing line segment = 4 + 6 = 10

- -

In this polygon to the left, there is one side missing.

Looking at the horizontal line segments, 20 and 15, we can determine what the top horizontal line segment is.
The missing line segment is 20 – 15 = $\boxed{5}$

Also, notice the vertical line segments, 3, 9, and 12. The two smaller vertical line segments must add to equal the larger vertical line segment.

- -

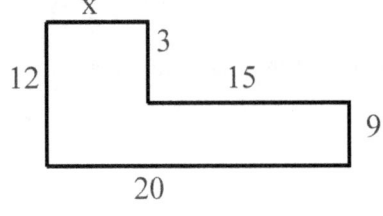

In this polygon to the left, there is one side missing.

Looking at the vertical line segments, 17 and 10, we can determine what the top vertical line segment is.
The missing line segment is 17 – 10 = $\boxed{7}$

Also, notice the horizontal line segments, 12, 14, and 26. The two smaller horizontal line segments must add to equal the larger horizontal line segment.

Your Turn Problem # 1

Find the missing side.

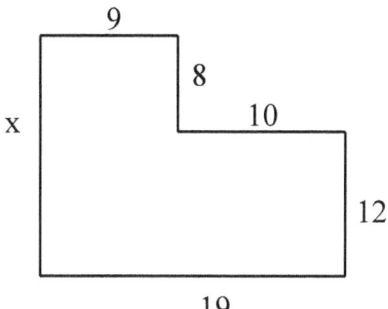

Answer:_____

Your Turn Problem # 2

Find the missing side. (labeled x)

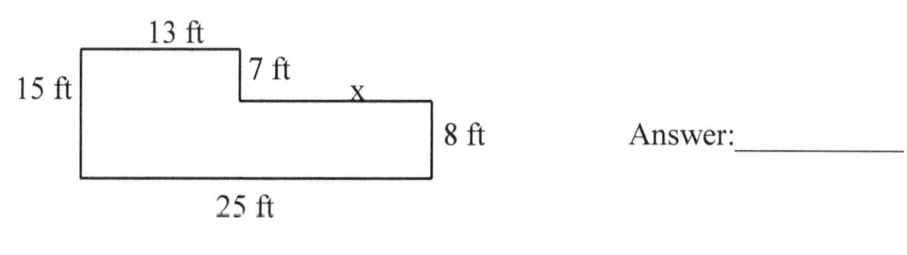

Answer:_____

Your Turn Problem # 3

Find the missing side. (labeled x)

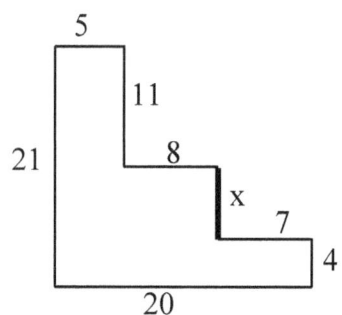

Answer:_____

Your Turn Problem # 4

Find the missing sides. (labeled x, y, and z)

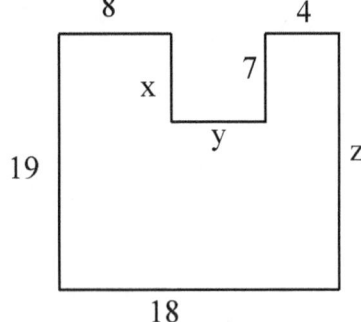

Answer:_____

Your Turn Problem # 5

Find the missing side. (labeled x and y)

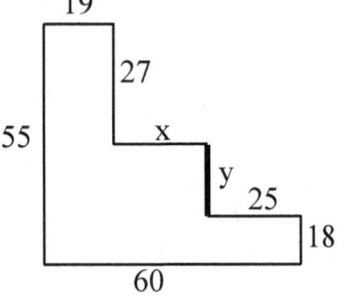

Answer:_____

Once we know the value of all of the sides, we can determine the perimeter by adding up all of the sides.

Example: Find the perimeter of the geometric figure.

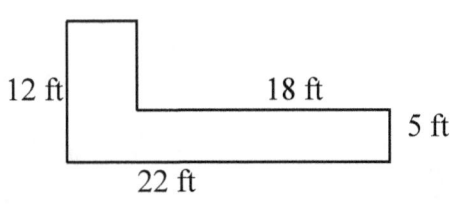

The horizontal missing segment is $22 - 18 = 4$.

The vertical missing segment is $12 - 5 = 7$.

To find the perimeter, add all of the sides.

$12 + 22 + 5 + 18 + 4 + 7 = \boxed{68 \text{ ft}}$

Your Turn Problem #6

Find the perimeter of the geometric figure.

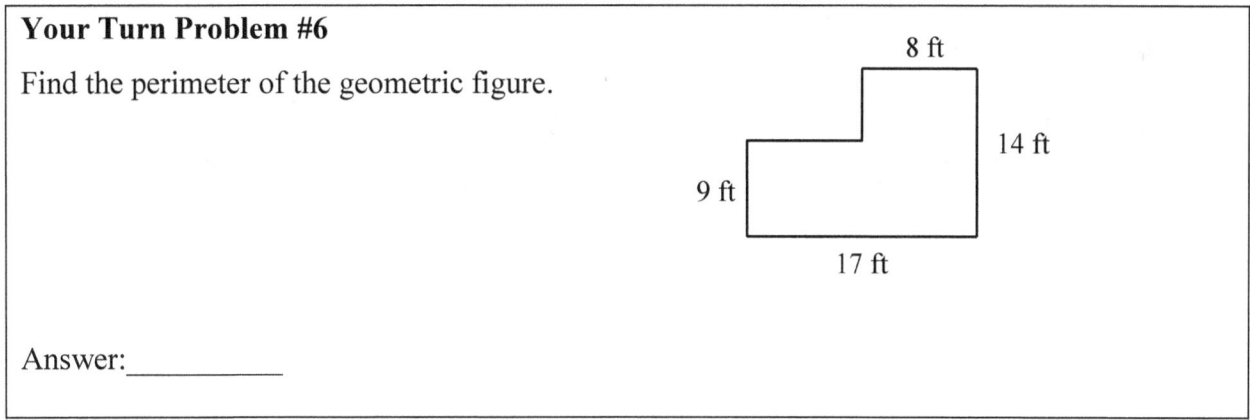

Answer:_____

Your Turn Problem # 7

Find the perimeter of the geometric figure.

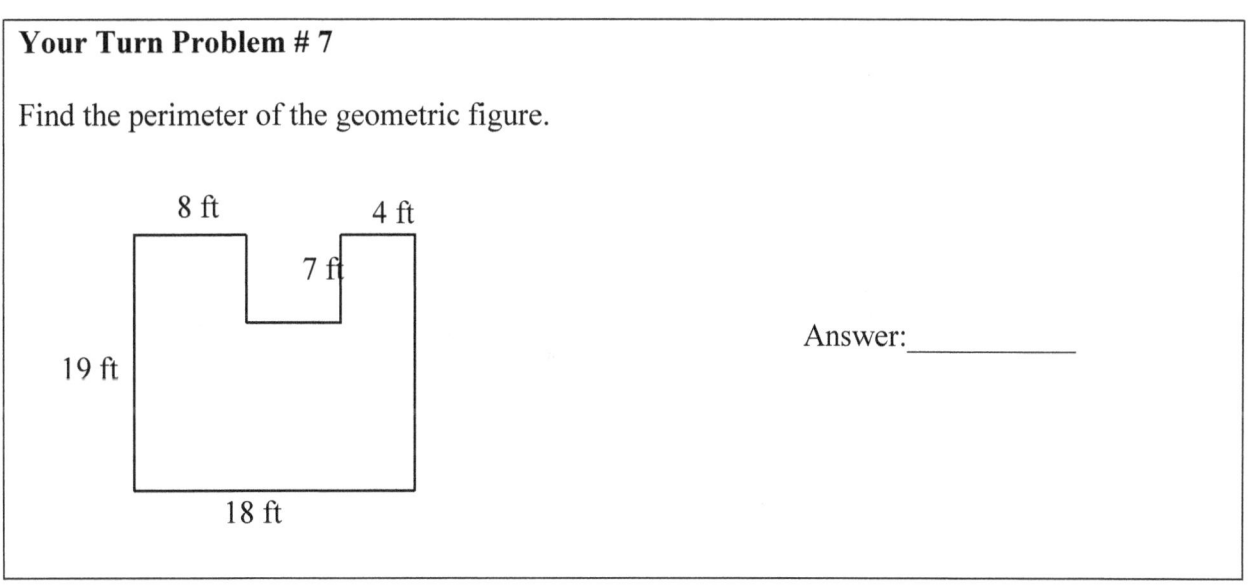

Answer:_____

3.3 Homework: Geometric Applications - Perimeter

Find the perimeter of each figure.

1.

(square)

4 ft

2.

3 yd

7 yd

3.

7 ft   4 ft

8 ft

4.

5 ft

10 ft

13 ft

4 ft

5.

18 ft   15 ft

6 ft

20 ft

6.

5 ft   2 ft

3 ft

4 ft

15 ft

7.

5 ft

6 ft

17 ft

7 ft

5 ft

16 ft

8.

3 m

15 m   8 m

9 m

4 m

14 m

3.3 Homework: Geometric Applications – Perimeter cont.

Find the perimeter of each figure.

9.

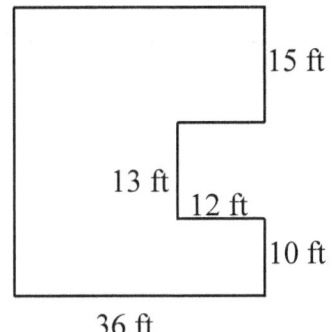

10.

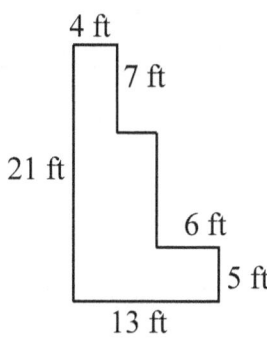

11. Find the perimeter of a rectangle if the length is 15 ft and the width is 7 ft.

12. Find the perimeter of a basketball court if it has dimensions of 50 ft by 94 ft.

13. Find the perimeter of a football field which is 120 yards long by 53 yards wide.

14. Find the perimeter of a triangle if the lengths of the sides are 17 ft, 12 ft and 27 ft.

15. Find the perimeter of a square if the length of a side is 19 ft.

3.4 Geometric Applications – Area and Volume

Area of a Rectangle

The area of a rectangle is length multiplied by width. Area usually measured in square units.

Examples: $1400\,\text{ft}^2$ (square feet), $5\,\text{mi}^2$ (square miles)

$$\boxed{\text{Area of a rectangle} = \text{length} \times \text{width}}$$

Example 1. Find the area of the rectangle.

5 ft

12 ft

Answer: The area of this rectangle is $12\ \text{ft} \times 5\ \text{ft} = 60\,\text{ft}^2$.

Your Turn Problem #1

Find the area of the rectangle.

14 m

11 m

Answer:_____

Area of a Parallelogram

A parallelogram is a 4-sided shape formed by two pairs of parallel lines. Opposite sides are equal in length and opposite angles are equal in measure. To find the area of a parallelogram, multiply the base by the height. The formula is:

$$\boxed{\text{Area of a parallelogram} = \text{base} \times \text{height}}$$

Note: parallel lines are lines that are sketched in the exact same direction.

Example of a parallelogram

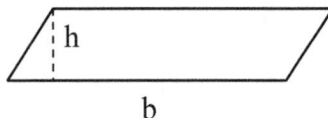

b

Example 2. Find the area of the parallelogram.

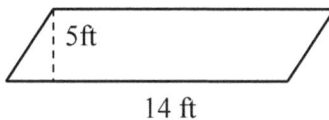

Answer: The area of this parallelogram is 14 ft × 5 ft = 70 ft² .

Your Turn Problem #2

Find the area of the parallelogram.

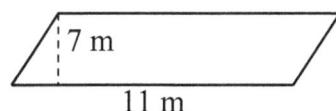

Answer:_____

Area of a Triangle

The area of a triangle is (base × height) ÷ 2

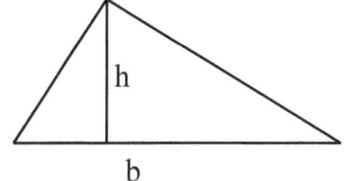

h=height , b=base

Note: The area of a triangle is half the area of a parallelogram. Therefore, if we only want half of the parallelogram, we divide the area of a parallelogram by 2.

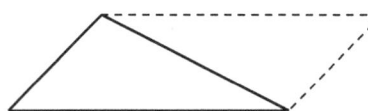

Area of a Parallelogram: $A = b \times h$
Area of a Triangle: $A = (b \times h) \div 2$

Example 3. Find the area of the triangle.

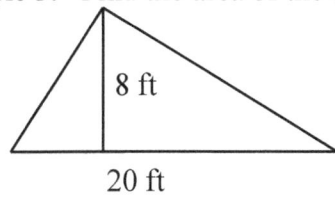

$$A = (20\,\text{ft} \times 8\,\text{ft}) \div 2$$

$$A = (160\,\text{ft}^2) \div 2$$

Answer: The area of this triangle $= 80\,\text{ft}^2$.

Your Turn Problem #3

Find the area of the triangle.

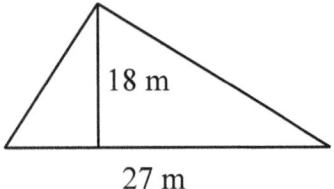

27 m

Answer:_____

The sum of rectangles

The area of some polygons can be found by separating the figure into rectangles.

By drawing a vertical line, we can make two rectangles. We can then find the area by finding the area of each individual rectangle A and B.

Example 4. Find the area of the geometric figure.

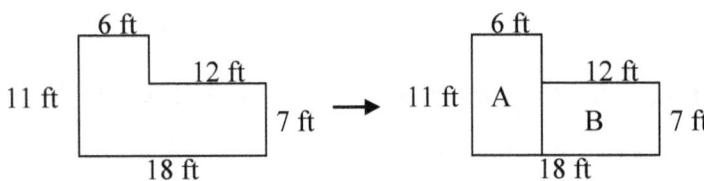

First, draw a line segment creating two rectangles A and B. (shown above)

To find the area of each rectangle, multiply the length and width

A: We have both length and width of rectangle A

Therefore, the area of rectangle A is 11 ft × 6 ft = 66 ft^2.

B: We have both the length and width of the rectangle B.

The width is 7 ft and the length is 12 feet.

Therefore, the area of rectangle B is 12 ft × 7 ft = 84 ft^2.

Answer: The total area of the figure is $66 \, \text{ft}^2 + 84 \, \text{ft}^2 = 150 \, \text{ft}^2$.

Your Turn Problem #4

Find the area of the geometric figure.

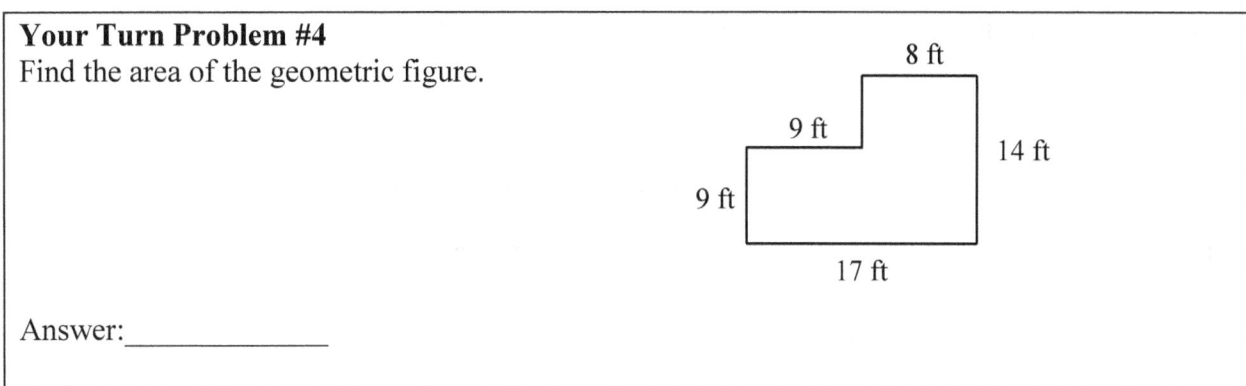

Answer:_____

Example 5. Find the area of the geometric figure.

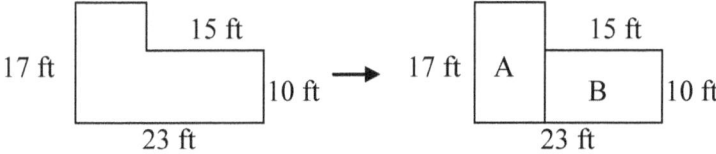

First, draw a line segment creating two rectangles A and B. (shown above)

To find the area of each rectangle, multiply the length and width.

A: We have the length of the rectangle A (17), but not the width.
23 is the length all the way from side to side. So subtract 15 to get the top horizontal line segment.
The width of rectangle A: 23 – 15 = 8.

Therefore, the area of rectangle A is 17 ft × 8 ft = 136 ft^2.

B: We have both length and width of rectangle B.

Therefore, the area of rectangle A is 10 ft × 15 ft = 150 ft^2.

Answer: The total area of the figure is $36\,\text{ft}^2 + 150\,\text{ft}^2 = 286\,\text{ft}^2.$

Your Turn Problem #5

Find the area of the geometric figure.

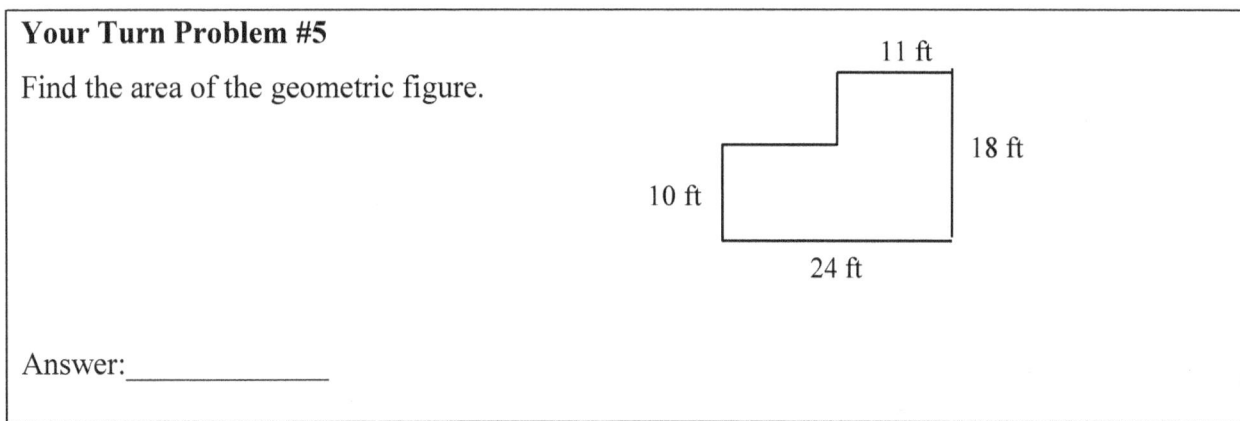

Answer:_____

Thus far, we have discussed perimeter and area.

Examples of the units for *perimeter* are: ft, m, mi, km, …

Examples of units for *area* usually have the squared symbol.: $ft^2, km^2, mi^2, …$

The area of a house is 1400 ft^2 (square feet). A fire has burned 500 mi^2 (square miles).

Volume of a rectangular solid.

Volume is how much a three-dimensional shape occupies. The volume of a container is generally understood to be the capacity of the container.

The formula for Volume of a rectangular solid is:

$$V = B \times H \times W$$

Where B = base, H = height, and W = width
The order does not matter because multiplication is commutative and associative.

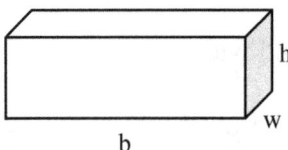

The units for volume will be cubic: ft^3, m^3, cm^3, cubic feet, cubic meters, cubic centimeters. Cubic centimeters are also called cc's.

Example 6. Find the volume of the rectangular solid.

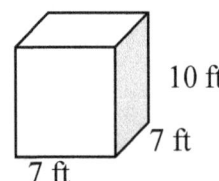

10 ft

7 ft

7 ft

Answer: The volume of this rectangular solid is 7 ft × 7 ft × 10 ft = $490 \, ft^3$.

Your Turn Problem #6

Find the volume of this rectangular solid.

9 ft

7 ft

15 ft

Answer:_____

3.4 Homework: Geometric Applications – Area and Volume

Find the area enclosed by the geometric figure.

1. 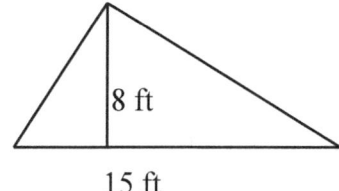 (square)

9 ft

2.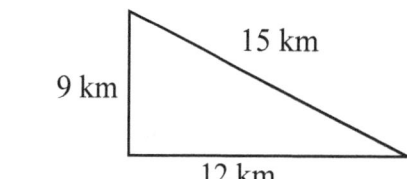

7 yd

12 yd

3.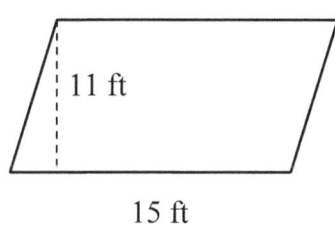

8 ft

15 ft

4.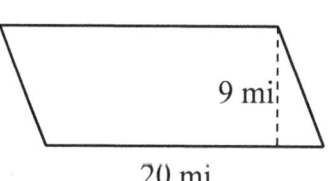

15 km

9 km

12 km

5.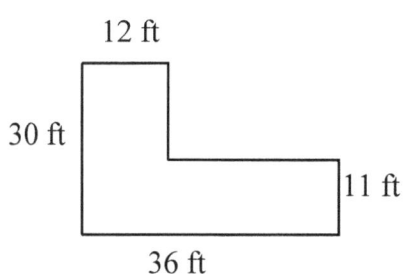

11 ft

15 ft

6.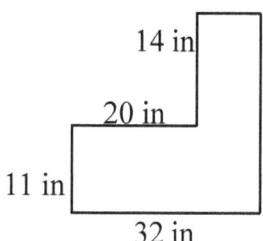

9 mi

20 mi

7.

12 ft

30 ft

11 ft

36 ft

8.

14 in

20 in

11 in

32 in

3.4 Homework: Geometric Applications – Area and Volume cont.

Find the volume enclosed by the geometric figure.

9.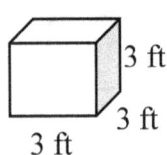
3 ft
3 ft
3 ft

10.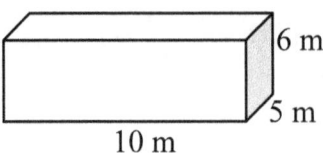
6 m
5 m
10 m

11. Find the area of a rectangle if the length is 18 feet and the width is 15 feet.

12. Find the area of a basketball court if it has dimensions of 50 ft by 94 ft.

13. Find the area of a triangle if the base is 34 feet and the height is 19 feet.

14. A room has an area of 400 square feet. A contractor has given you a bid to tile the room for $9 per square foot. How much will it cost to tile the room?

15. A room measures 28 ft by 16 feet. If a contractor gives you a bid to tile the room for $7 per square foot, how much will it cost to tile the room?

16. A room measures 24 ft by 12 feet. If a contractor gives you a bid to tile the room for $11 per square foot, how much will it cost to tile the room?

3.5 Divisibility Rules

Divisibility

Divisible: A number is said to be "divisible" by a smaller number if the smaller will divide into the larger "evenly" with no remainder.

Example: 36 is divisible by 2 because there is no remainder.

$$\begin{array}{r} 18 \\ 2\overline{)36} \\ -2 \\ \hline 16 \\ -16 \\ \hline 0 \end{array}$$

Divisibility Tests

A *test of divisibility* is a procedure for determining whether a number is divisible by another number by performing a "test" rather that actually dividing.

Divisibility Test for 2

All even numbers are divisible by 2. Even numbers end with either of the digits: 2, 4, 6, 8, or 0.

> **A number is divisible by 2 if it has a ones digit of 0, 2, 4, 6, or 8 (that is, it has an even ones digit).**

Example 1. a)   Is 584 divisible by 2?

Answer: Yes, because the last digit is an even number.

To find the quotient, we would still need to divide, but we know there is no remainder.

Example 1. b)   Is 39 divisible by 2?

Answer: No, because the last digit is an odd number. If we divided 2 into 39, there would be a remainder.

Your Turn Problem #1

Determine whether the following numbers are divisible by 2. (State yes or no and why.)

a) 90 :_____

b) 243:_____

c) 2,518:_____

Divisibility Test for 3

Let's make a multiplication fact table for 3.

$3 \times 0 = 0$	$3 \times 7 = 21$
$3 \times 1 = 3$	$3 \times 8 = 24$
$3 \times 2 = 6$	$3 \times 9 = 27$
$3 \times 3 = 9$	$3 \times 10 = 30$
$3 \times 4 = 12$	$3 \times 11 = 33$
$3 \times 5 = 15$	$3 \times 12 = 36$
$3 \times 6 = 18$	$3 \times 13 = 39$

Notice the patterns.

1. Each result increases by 3.

2. The sum of the digits is: 3, 6, 9 or 12. $(3 \times 6 = 18 , 1 + 8 = 9)$

    $(3 \times 8 = 24 , 2 + 4 = 6)$

If we kept listing more products with a factor of 3, we would also find that the sum of its digits may be a 15, 18, 21, etc. This gives us the following divisibility test.

A number is divisible by 3 if the sum of its digits is divisible by 3.

Example 2a.   Is 57 divisible by 3?

Answer: Yes, 57 is divisible by 3 because 5 + 7 = 12, and 12 is divisible by 3. If we wanted to know how many times 3 divides into 57, we would actually have to divide. This test only tells us that 3 will divide evenly into 57.

$$
\begin{array}{r}
19 \\
3\overline{)57} \\
-3 \\
\hline
27 \\
-27 \\
\hline
0
\end{array}
$$

This means that $3 \times 19 = 57$.

Example 2b.   Is 621 divisible by 3?

Answer: Yes, 621 is divisible by 3 because 6+2+1=9, and 9 is divisible by 3.

Example 2c.   Is 443 divisible by 3?

Answer: No, 443 is not divisible by 3 because 4+4+3=11, and 11 is not divisible by 3.

If we tried to divide 3 into 443, there would be a remainder.

Your Turn Problem #2
Determine whether the following numbers are divisible by 3. (State yes or no and why.)

a) 171:_____

b) 297:_____

c) 613:_____

Divisibility Test for 5

Let's make a multiplication fact table for 5.

$5 \times 0 = 0$	$5 \times 7 = 35$
$5 \times 1 = 5$	$5 \times 8 = 40$
$5 \times 2 = 10$	$5 \times 9 = 45$
$5 \times 3 = 15$	$5 \times 10 = 50$
$5 \times 4 = 20$	$5 \times 11 = 55$
$5 \times 5 = 25$	$5 \times 12 = 60$
$5 \times 6 = 30$	$5 \times 13 = 65$

Notice the pattern. Of course each result increases by 5, but also the last digit is either a 0 or a 5.

A number is divisible by 5 if its last digit is either a 5 or 0.

Example 3a. Is 95 divisible by 5?

Answer: Yes, 95 is divisible by 5 because the last digit is a 5.

Example 3b. Is 553 divisible by 5?

Answer: No, 553 is not divisible by 5 because the last digit is not a 0 or a 5. It is a 3.

Example 3c. Is 130 divisible by 3?

Answer: Yes, 130 is divisible by 5 because the last digit is a 0.

Your Turn Problem #3
Determine whether the following numbers are divisible by 5. (State yes or no and why.)

a) 230:_____

b) 175:_____

c) 352:_____

Divisibility Test for 9

Let's make a multiplication fact table for 9.

$9 \times 1 = 9$

$9 \times 2 = 18$

$9 \times 3 = 27$

$9 \times 4 = 36$

$9 \times 5 = 45$

$9 \times 6 = 54$

$9 \times 7 = 63$

$9 \times 8 = 72$

$9 \times 9 = 81$

$9 \times 10 = 90$

Notice the patterns.

1. Each result increases by 9.

2. The first digit increases by 1 and the last digit decreases by 1.

3. Each result has a match in reverse order, i.e., 18 and 81, 27 and 72.

4. The sum of the digits equals 9. $(1 + 8 = 9, \ 4 + 5 = 9)$

Lets list a few more to observe the pattern.

$9 \times 11 = 99$

$9 \times 12 = 108$

$9 \times 13 = 117$

$9 \times 14 = 126$

$9 \times 32 = 288$

$9 \times 73 = 657$

The sum of the digits will not always equal 9. However, it will be a number that 9 divides evenly into, such as 18, 27 or 36.

A number is divisible by 9 if the sum of its digits is divisible by 9 (i.e., 9, 18, 27, 36, …).

Example 4a.   Is 441 divisible by 9?

Answer: Yes, 441 is divisible by 9 because the sum of its digits is divisible by 9.

(4 + 4 + 1 = 9 and 9 is divisible by 9.)

Example 4b.   Is 8937 divisible by 9?

Answer: Yes, 8937 is divisible by 9 because the sum of its digits is divisible by 9.

(8 + 9 + 3 +7 = 27 and 27 is divisible by 9.)

Example 4c.   Is 229 divisible by 9?

Answer: No, 229 is not divisible by 9 because the sum of its digits is not divisible by 9.

(2 + 2 + 9 = 13 and 13 is not divisible by 9.)

Your Turn Problem #4

Determine whether the following numbers are divisible by 9. (State yes or no and why.)

a) 5481:_____

b) 329:_____

c) 504:_____

Divisibility Test for 10

Let's make a multiplication fact table for 10.

$10 \times 0 = 0$	$10 \times 7 = 70$
$10 \times 1 = 10$	$10 \times 8 = 80$
$10 \times 2 = 20$	$10 \times 9 = 90$
$10 \times 3 = 30$	$10 \times 10 = 100$
$10 \times 4 = 40$	$10 \times 11 = 110$
$10 \times 5 = 50$	$10 \times 12 = 120$
$10 \times 6 = 60$	$10 \times 13 = 130$

Notice the pattern. Of course each result increases by 10, but also the last digit is always a 0.

A number is divisible by 10 if its last digit is 0.

Example 5a.   Is 310 divisible by 10?

Answer: Yes, 310 is divisible by 10 because the last digit is a 10.

Example 5b.   Is 105 divisible by 10?

Answer: No, 105 is not divisible by 10 because the last digit is not a 0. It is a 5.

Your Turn Problem #5

Determine whether the following numbers are divisible by 10. (State yes or no and why.)

a) 540:_____

b) 300:_____

c) 255:_____

Thus far, we have only discussed divisibility tests for 2, 3, 5, 9 and 10. There are certainly more divisibility rules. Memorizing more divisibility tests may be more complicated than simply dividing to check divisibility.

Divisibility for 7

There is a divisibility test for 7 however it is easier to just divide by 7.

Example 6a.   Is 91 divisible by 7?

Divide 7 into 91 to determine if 91 is divisible by 7.

$$
\begin{array}{r}
13 \\
7\overline{)91} \\
-7 \\
\hline
21 \\
-21 \\
\hline
0
\end{array}
$$

Answer: Yes, 91 is divisible by 7. $7 \times 13 = 91$

Example 6b.   Is 107 divisible by 7?

Answer: No, 107 is not divisible by 7 because it doesn't divide evenly into 107.

$$
\begin{array}{r}
15 \\
7\overline{)107} \\
-7 \\
\hline
39 \\
-35 \\
\hline
4
\end{array}
$$

Example 6c.   Is 119 divisible by 7?

Divide 7 into 119 to determine if 119 is divisible by 7.

$$
\begin{array}{r}
17 \\
7\overline{)119} \\
-7 \\
\hline
49 \\
-49 \\
\hline
0
\end{array}
$$

Answer: Yes, 119 is divisible by 7. $7 \times 17 = 119$

Your Turn Problem #6

Determine whether the following numbers are divisible by 7. If it is divisible by 7, show the factors which multiply to equal the given number.

a) 77:_____

b) 161:_____

c) 131:_____

Example 7. Determine which of the numbers 2, 3, 5 and 10 will divide evenly into 225?

Solution: Is 225 divisible by 2? Answer: No, 225 doesn't end with an even digit.

Is 225 divisible by 3? Answer: Yes, 2 + 2+ 5 =9 and 9 is divisible by 3.

Is 225 divisible by 5? Answer: Yes, 225 ends with a 0 or a 5.

Is 225 divisible by 10? Answer: No, 225 doesn't end with a 0.

Answer: 225 is divisible by only 3 and 5.

Your Turn Problem #7
Determine which of the numbers 2, 3, 5 and 10 will divide evenly into 710?

Is 710 divisible by 2? Answer:_____

Is 710 divisible by 3? Answer:_____

Is 710 divisible by 5? Answer:_____

Is 710 divisible by 10? Answer:_____

Answer: 710 is divisible by:_____.

3.5 Homework: Divisibility Rules

Determine which of the numbers 2, 3, 5 and 10 will divide exactly into each of the following numbers.

1. 540:_____

    2:

    3:

    5:

    10:

2. 346:_____

    2:

    3:

    5:

    10:

3. 621:_____

    2:

    3:

    5:

    10:

4. 2,690:_____

    2:

    3:

    5:

    10:

5. 5,211:_____

    2:

    3:

    5:

    10:

6. 4,002:_____

    2:

    3:

    5:

    10:

3.5 Homework: Divisibility Rules cont.

Determine which of the numbers 2, 3, 5 and 10 will divide exactly into each of the following numbers.

7.   6,732:_____

    2:

    3:

    5:

    10:

8.   9,017:_____

    2:

    3:

    5:

    10:

9.   10,950:_____

    2:

    3:

    5:

    10:

10. 12,579:_____

    2:

    3:

    5:

    10:

Determine if 7 will divide into the following numbers. If it is divisible by 7, show the factors which multiply to equal the given number.

11. 91:_____

12. 157:_____

13. 133:_____

14. 77:_____

15. 1,463:_____

3.5 Homework: Divisibility Rules cont.

Use the following list for problems 16 through 20.
45,   72,   158,   260,   378,   569,   570,   585,   3,541,   4,530,   8,300

16. Which numbers are divisible by 2?_____

17. Which numbers are divisible by 3?_____

18. Which numbers are divisible by 5?_____

19. Which numbers are divisible by 10?_____

20. Which numbers are divisible by 9?_____

3.6 Prime Numbers and Composite Numbers

Numbers that are multiplied are called factors. For example, $3 \cdot 4 = 12$.

The numbers 3 and 4 are called **Factors**. The result, 12, is called the product.

We say that 3 and 4 are factors of 12.

Prime Numbers (sometimes just called "Primes")

A prime number is a natural number that has only two distinct factors which are 1 and itself. An example of a prime number is 7. The only factors of 7 are 1 and 7. We could also say that no other number besides 1 and itself divides evenly into a prime number.

A number such as 12 is not a prime number since it has more than two distinct factors.

$3 \cdot 4 = 12$, and $2 \cdot 6 = 12$

List of the first ten prime numbers.
2, 3, 5, 7, 11, 13, 17, 19, 23, 29, etc.

Note that 1 is not on the list. Since a prime number has two different factors, it does not qualify. So the number 1 is not a prime number.

Composite Numbers

Composite numbers are natural numbers with more than two different factors.

A composite number has factors other than 1 and itself.

Example: 12 is a composite number because $3 \cdot 4 = 12$. The factors of 12 are 1, 2, 3, 4, 6, and 12.

List of the first ten composite numbers.
4, 6, 8, 9, 10, 12, 14, 15, 16, 18, etc.

Note that 1 is not on the list. Since a composite number has more than two different factors, it does not qualify. So the number 1 is not a composite number. The number 1 is neither prime nor composite.

Determining if a number is a Prime Number.

A number is not a prime number if it is divisible by a number other than one and itself.

Procedure:

1. Use the divisibility to determine if the number is divisible by 2, 3 or 5.

2. If the number is not divisible by 2, 3 or 5, continue with the prime numbers beginning at 7.

3. Only continue this process until the prime number you are dividing with is a number that when multiplied by itself, the result is larger than the number being tested. Another way of knowing when to stop is "when the number in the quotient is smaller than the divisor."

If we **cannot** find a prime number that divides evenly into the number being tested, it is **prime**.

If we **can** find a prime number that divides evenly into the number being tested, it is **composite**.

Example 1. Is 411 a prime or composite number? (state reason)

Solution: Check sequentially the numbers 2, 3, and 5

Is 411 divisible by 2? No, doesn't end with an even digit.

Is 411 divisible by 3? Yes, since 4+1+1 = 6 which is divisible by 3.

Answer: 411 is a composite number because it is divisible by 3. $(3 \cdot 137 = 411)$

Your Turn Problem #1

Is 237 a prime or composite number? State the reason.

Example 2. Is 145 a prime or composite number? (state reason)

Solution: Check sequentially the numbers 2, 3, and 5

Is 145 divisible by 2? No, doesn't end with an even digit.

Is 145 divisible by 3? No, since 1+4+5 = 10 which is not divisible by 3.

Is 145 divisible by 5? Yes since the last digit ends with a 0 or 5.

Answer: 145 is a composite number because it is divisible by 5. ($5 \cdot 29 = 145$)

Your Turn Problem #2

Is 135 a prime or composite number? State the reason.

Example 3. Is 131 a prime or composite number? (State the reason.)

Solution: Check sequentially the prime numbers.

Is 131 divisible by 2? No, doesn't end with an even digit.

Is 131 divisible by 3? No, since 1+3+1 = 5 which is not divisible by 3.

Is 131 divisible by 5? No, since the last digit does not end with a 5.

Is 131 divisible by 7? No, there is a remainder.

$$\begin{array}{r} 18 \\ 7\overline{)131} \\ -7 \\ \hline 61 \\ -56 \\ \hline 5 \end{array}$$

Is 131 divisible by 11? No, there is a remainder.

$$\begin{array}{r} 11 \\ 11\overline{)131} \\ -11 \\ \hline 21 \\ -11 \\ \hline 10 \end{array}$$

Is 131 divisible by 13? If we multiply 13 by itself, the result is larger than 131.
$13 \times 13 = 169$ which is larger than 131. So we don't need to try 13. If you divide the 13 into 131, the number in the quotient will be smaller than the 13.

Answer: 131 is a prime number because it is only divisible by 1 and itself.

Your Turn Problem #3

Is 149 a prime or composite number? State the reason.

Multiples

A multiple of a natural number is a product of that number and any natural number. In other words, take the number and multiply it by 1, 2, 3, …

Example: Multiples of 6: 6, 12, 18, 24, 30, 36, …

(6×1=6,   6×2=12,   6×3=18,   6×4=24,   6×5=30, etc.)

There is a relationship between prime numbers, composite numbers and multiples. That is, every composite number is a multiple of a prime number.

Example 4a. Complete the list for the first 10 multiples of 2.

2 , 4 , 6 , ____ , ____ , ____ , ____ , ____ , ____ , ____

Each number above, except the 2, is a composite number.

Example 4b. Complete the list for the first 10 multiples of 13.

13 , 26 , 39 , ____ , ____ , 78 , ____ , 104 , 117 , ____

Each number above, except the 13, is a composite number.

Example 4c. Complete the list for the first 10 multiples of 9.

9 , 18 , 27 , ____ , ____ , ____ , ____ , ____ , ____ , ____

Each number above, is a composite number (9 is not prime).

3.6 Homework: Prime Numbers and Composite Numbers

1. There are 25 prime numbers less than 100. List all of them.

___ , ___ , ___ , ___ , ___ , ___ , ___ , ___ , ___ , ___ , ___ , ___ , ___ , ___ , ___ ,

___ , ___ , ___ , ___ , ___ , ___ , ___ , ___ , ___ , ___

Identify each number as prime or composite. If composite, show why (what prime number is it divisible by).

2. 11:_____

3. 57:_____

4. 23:_____

5. 141:_____

6. 18:_____

7. 47:_____

8. 91:_____

9. 111:_____

10. 235:_____

11. 152:_____

12. List the first 10 multiples of 9: ___ , ___ , ___ , ___ , ___ , ___ , ___ , ___ , ___ , ___.

13. List the first 10 multiples of 25: ___ , ___ , ___ , ___ , ___ , ___ , ___ , ___ , ___ , ___.

14. List the first 10 multiples of 11: ___ , ___ , ___ , ___ , ___ , ___ , ___ , ___ , ___ , ___

3.7 Prime Factorization and Factors

Prime factorization

Prime factorization is the process of rewriting a composite number as a *product of only primes*. Prime factorization can only be performed on composite numbers.

Examples: $24 = 2 \cdot 2 \cdot 2 \cdot 3$ or $2^3 \cdot 3$

$35 = 5 \cdot 7$

$44 = 2 \cdot 2 \cdot 11$ or $2^2 \cdot 11$

Note: The product of the prime factorization must equal the composite number.

Determining the Prime Factorization of a Composite Number

There are two techniques for finding the prime factorization of a composite number. Please use the method you feel most comfortable with.

Procedure: Tree Method for Determining the Prime Factorization of a Composite Number

Find two factors whose product is the original number. Draw two branches under the original number and write each factor under each branch. Continue the process until the numbers at below each branch is a prime number.

Example 1: Find the prime factorization 24.

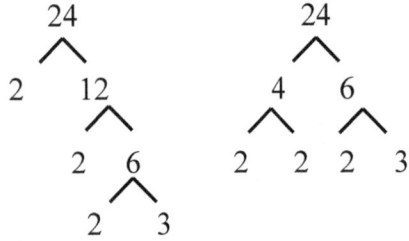

The Prime Factorization of 24 is $2 \cdot 2 \cdot 2 \cdot 3$ or $2^3 \cdot 3$

Note: It does not matter which two factors we start with (just not 1 and the number given). The result will be the same.

Procedure: Division Method for Determining the Prime Factorization of a Composite Number

Divide the given number by the smallest prime number that divides evenly into the number. Then, directly above, divide the result (answer) by the smallest prime number that divides evenly in that number. Continue this process until you have an answer which is a prime number. The prime numbers obtained is the prime factorization.

Example 1 (again):   Find the prime factorization 24.

1. 2 is the smallest prime that divides evenly into 24.

2. Smallest prime that divides evenly into 12: 2 again.

3. Smallest prime that divides evenly into 6: still 2.

4. The top number is prime, so we're done.

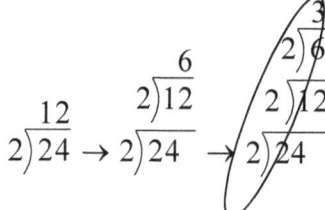

The Prime Factorization of 24 is $2 \cdot 2 \cdot 2 \cdot 3$ or $2^3 \cdot 3$

Your Turn Problem #1

Find the prime factorization 60.

Answer:_____

Listing all of the Factors for a Composite Number

We say that 3 and 4 are factors of 12.

However, the number 12 has other factors: $2 \cdot 6 = 12$, and $1 \cdot 12 = 12$.

So the complete list of factors for 12 is 1, 2, 3, 4, 6, and 12.

Another way of stating what factors is: "Numbers that divide evenly into a given number".

1, 2, 3, 4, 6, and 12 all divide evenly into 12. The number 5 is not a factor because it does not divide evenly into 12 (there is a remainder).

Once again, there are two methods for listing all of the factors of a given number.

Procedure for Listing all of the Factors for a Composite Number by Checking Divisibility:

Start by listing the product of any given number, 1 and itself. Continue listing products sequentially beginning with the numbers 2, 3, 4 and so on by determining if it is a factor. When the number you are determining if it is a factor is already listed in the factors, the list is complete.

Example 2. List *all* the factors of 24.

Check sequentially the numbers 1, 2, 3, and so on, to see if we can form any factorizations.

1×24: is the first product. 2 is a factor because the last digit is even. Divide 2 into 24 to find the product (2×12). 3 divides evenly into 24 because the sum of its digits is 6, and 3 divides evenly into 6 (3×8). 4 divides evenly into 24 (4×6). 5 does not divide evenly into 24. 6 does divide evenly into 24, but it was already found.

$$\frac{24}{}$$

$1 \cdot 24$

$2 \cdot 12$

$3 \cdot 8$ **Answer:** The factors of 24 are 1, 2, 3, 4, 6, 8, 12, and 24.

$4 \cdot 6$

Your Turn Problem #2

List all the factors of 18.

Answer:_____

Finding all Factors of a Composite Number using the Prime Factorization

The previous method is definitely a little easier but, only for smaller numbers. This method will help us out with the larger numbers. Once we have the prime factorization of number, we can use it to find the factors of the number. To be a factor:, it must be: 1, the number itself, one of the prime factors, or the product of 2 or more of the prime factors.

Example: The Prime Factorization of 24 is $2 \cdot 2 \cdot 2 \cdot 3$

1 is always a factor. **24** the number itself is a factor. The prime numbers are **2** and **3** which are factors. Then the product of 2 or more of the prime factors:

$2 \times 2 = \mathbf{4}, \quad 2 \times 3 = \mathbf{6}, \quad 2 \times 2 \times 2 = \mathbf{8}, \quad 2 \times 2 \times 3 = \mathbf{12}$

Using this method, we have obtained 1, 2, 3, 4, 6, 8, 12 and 24

Procedure for Listing all of the Factors for a Composite Number using Prime Factorization

1. Find the prime factorization.

2. 1 and the number itself are factors.

3. Each prime number in the prime factorization is a factor.

4. Products found by all combinations of the prime factors are factors.

Example 3. List *all* the factors of 135.

Solution:

1. Find the prime factorization (use either method). $135 = 3 \cdot 3 \cdot 3 \cdot 5$

2. 1 and 135 are factors.

3. 3 and 5 are factors (each prime number in prime factorization is a factor).

4. Find all products by combinations of 3, 3, 3 and 5.

$$3 \times 3 = 9, \quad 3 \times 5 = 15, \quad 3 \times 3 \times 3 = 27, \quad 3 \times 3 \times 5 = 45$$

$$\begin{array}{r} 5 \\ 3\overline{)15} \\ 3\overline{)45} \\ 3\overline{)135} \end{array}$$

Note: we do not need to multiply all of the numbers. That would equal 135 which is already listed.

Answer: All of the factors of 135 are 1, 3, 5, 9, 15, 27, 45 and 135.

Your Turn Problem #3

List all the factors of 80.

Answer:_____

Prime Factorization and Factors of a Prime Number

If asked to find the prime factorization of a number which is actually a prime number, simply state the number is a prime number.

If asked to find the factors of a prime number, the factors are only 1 and itself.

Example: The factors of 11 are only 1 and 11.

3.7 Homework: Prime Factorization and Factors

Find the prime factorization of each number.

1.   36 = _____

2.   24 = _____

3.   360 = _____

4.   48 = _____

5.   100 = _____

6.   65 = _____

7.   13 = _____

8.   55 = _____

3.7 Homework: Prime Factorization and Factors

Find the prime factorization of each number.

9. $240 = $ _____

10. $144 = $ _____

11. $147 = $ _____

12. $175 = $ _____

13. $195 = $ _____

14. $154 = $ _____

15. $875 = $ _____

16. $504 = $ _____

3.7 Homework: Prime Factorization and Factors cont.

List all the factors of each of the following numbers

17. 36 =_____

18. 24 =_____

19. 15 =_____

20. 48 =_____

21. 100 =_____

22. 65 =_____

23. 13 =_____

24. 55 =_____

3.7 Homework: Prime Factorization and Factors cont.

List all the factors of each of the following numbers

25. 64 =_____

26. 6 =_____

27. 90 =_____

28. 60 =_____

29. 30 =_____

30. 56 =_____

3.8 The GCF (Greatest Common Factor)

Greatest Common Factor (GCF)

The greatest common factor of two or more numbers is the largest factor that the numbers have in common. Another way of stating the GCF is: the largest number that divides evenly into the numbers given.

For example: Find the GCF of 30 and 45.

Solution: We could list all of the factors of each number. Then the GCF would be the largest number on both lists.

30 = 1, 2, 3, 5, 6, 10, 15, 10

45 = 1, 3, 5, 9, 15, 45

Therefore the **GCF is 15** because it is the largest number common to both lists of factors.

We also stated that the GCF is the largest number that divides evenly into both numbers. Since 30 and 45 are divisible by 3, 5, and 15 and the greatest of those numbers is 15, then the **GCF = 15**.

Instead of listing all of the factors, or trying to come up with the numbers that divide into the numbers given, we have a better technique for finding the GCF using Prime Factorization.

Procedure for finding the GCF using Prime Factorization

Step 1. Find the prime factorization of each number. If the number is prime, simply write down the same prime number after the "=" sign.

Step 2. Circle any prime number that appears in all prime factorization lists. It is possible that two or more of any prime number can be common to all rows.

Step 3. Multiply the common primes from one row together. This is the GCF. If there is only one prime factor in common, then there is nothing to multiply and that prime factor is the GCF.

If there are no primes in common, then the GCF is 1. (1 is always a factor of any number.)

Example 1. Find the GCF of 12 and 18.

Step 1. Find the prime factorization of each number.

$$12 = 2 \cdot 2 \cdot 3$$
$$18 = 2 \cdot 3 \cdot 3$$

Step 2. Circle any prime that appears in both rows.

Step 3. Multiply the common primes together from one row.

Answer: The GCF = $2 \cdot 3 = 6$.

Prime Factorization

$$\begin{array}{r} 3 \\ 2\overline{)6} \\ 2\overline{)12} \end{array} \qquad \begin{array}{r} 3 \\ 3\overline{)9} \\ 2\overline{)18} \end{array}$$

(Tree method is also fine.)

Your Turn Problem #1

Find the GCF of 28 and 42.

$28 =$

$42 =$

$GCF =$

Example 2. Find the GCF of 54, 90 and 108.

Step 1. Find the prime factorization of each number.

$$54 = 2 \cdot 3 \cdot 3 \cdot 3$$
$$90 = 2 \cdot 3 \cdot 3 \cdot 5$$
$$108 = 2 \cdot 2 \cdot 3 \cdot 3 \cdot 3$$

Step 2. Circle any prime that appears in all three rows.

Step 3. Multiply the common primes together from one row.

Answer: The GCF = $2 \cdot 3 \cdot 3 = 18$.

Prime Factorization

$$\begin{array}{r} 3 \\ 3\overline{)9} \\ 3\overline{)27} \\ 2\overline{)54} \end{array} \qquad \begin{array}{r} 5 \\ 3\overline{)15} \\ 3\overline{)45} \\ 2\overline{)90} \end{array} \qquad \begin{array}{r} 3 \\ 3\overline{)9} \\ 3\overline{)27} \\ 2\overline{)54} \\ 2\overline{)108} \end{array}$$

Your Turn Problem #2

Find the GCF of 88, 132 and 220.

$88 =$

$132 =$

$220 =$

$GCF =$

Example 3. Find the GCF of 8 and 15.

Step 1. Find the prime factorization of each number.

$$8 = 2 \cdot 2 \cdot 2$$
$$15 = 3 \cdot 5$$

Prime Factorization

$$2\overline{)6}^{\,2}$$
$$2\overline{)12} \qquad 3\overline{)15}^{\,5}$$

Step 2. Circle any prime that appears in all three rows. (none)

Step 3. If there are no primes in common, then the GCF is 1.

Answer: The GCF = 1.

Your Turn Problem #3

Find the GCF of 9 and 25.

$9 =$

$25 =$

$GCF =$

Example 4. Find the GCF of 11, 22 and 44.

Step 1. Find the prime factorization of each number.

$$11 = 11$$
$$22 = 2 \cdot 11$$
$$44 = 2 \cdot 2 \cdot 11$$

Prime Factorization

$$2\overline{)22}^{\,11}$$
$$2\overline{)22} \qquad 2\overline{)44}^{\,11}$$

Step 2. Circle any prime that appears in all three rows.

Step 3. If there is only one prime factor in common, then there is nothing to multiply and that prime factor is the GCF.

Answer: The GCF = 11.

Your Turn Problem #4

Find the GCF of 13, 39 and 52.

$13 =$

$39 =$

$52 =$

$GCF =$

3.8 Homework: The GCF (Greatest Common Factor)

Find the greatest common factor (GCF) for each of the following groups of numbers

1.   12 and 18

   12 =

   18 =

   GCF =

2.   15 and 25

   15 =

   25 =

   GCF =

3.   22 and 14

4.   10 and 30

5.   21 and 28

6.   12, 36 and 60

3.8 Homework: The GCF (Greatest Common Factor) cont.

Find the greatest common factor (GCF) for each of the following groups of numbers

7.   26, 39, and 52

8.   55, 66, and 77

9. 25, 75, and 150

10. 12 and 25

11. 92 and 138

12. 38 and 57

3.9 The LCM (Least Common Multiple)

Multiples

A multiple of a natural number is a product of that number and any natural number.

In other words, take the number and multiply it by 1, 2, 3, …

Example: Multiples of 6: 6, 12, 18, 24, 30, 36, …

(6×1=6, 6×2=12, 6×3=18, 6×4=24, 6×5=30, etc.)

Least Common Multiple

The least common multiple, or LCM, of two natural numbers is the smallest number that is a multiple of both numbers.

Example: Observe the list of multiples for the following numbers.

6: 6, 12, 18, 24, 30, 36, 42, 48, 54, 60, 66, 72, 78, …

8: 8, 16, 24, 32, 40, 48, 56, 64, 72, 80, …

The **LCM** is the smallest number that appears in both lists. (Least Common Multiple)

24 is the smallest number that appears in both lists, therefore it is the LCM. There are other common multiples such as 48 and 72, but 24 is the lowest common multiple.

Finding a LCM is part of the process to adding and subtracting fractions with unlike denominators so it is very important. Writing down a list of multiples is sufficient method as long as the numbers are relatively small. What if the numbers are 84 and 90? Then, making a list of multiples is a little tedious. Fortunately, we have a better method.

Prime Factorization Method for Finding the LCM (Least Common Multiple)

Procedure: Finding the LCM using the Prime Factorization Method

Step 1. Find the prime factorization of each number. You can use either method. Use either the prime factorization tree or the division method.

Step 2. Find the prime factors that appear in either factorization.

Step 3. The LCM = product of these prime factors, writing each prime factor the greatest number of times that it occurs in any one of the prime factorizations.

Restating Step 3 from the procedure.

Start with the smallest prime number in either factorization.

Ask the following questions to yourself: "Which has the most of that prime number?" and "How many times does it occur in that row?" Write that prime number that same number of times. Then continue on with the next smallest prime number.

For example: Suppose the top row has three 7's and the bottom row has five 7's.

Question: Which row has the most? Answer: The bottom row. How many? It has five. The LCM will have five 7's.

What if both rows have the same amount of a prime number? Then choose the number of primes from the top row (same as the number of primes from the bottom row).

Common Mistake: Do not take the same prime factor from two different rows. Your LCM will be too big.

Example 1. Find the LCM of 15 and 24.

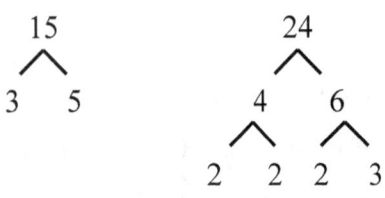

$$15 = 3 \cdot 5$$
$$24 = 2 \cdot 2 \cdot 2 \cdot 3$$
$$\text{LCM} = 2 \cdot 2 \cdot 2 \cdot 3 \cdot 5$$
$$= \boxed{120}$$

Step 1. Find the prime factorization of each number.

Step 2. Write "LCM=" below the prime factorization of each number.

Consider each prime number. Write it the greatest number of times it occurs in any one factorization.

2: it occurs three times in the 24.

3: it occurs once.

5: it occurs once.

Now multiply. This is the LCM.

Your Turn Problem #1

Find the LCM of 35 and 90.

$$35 =$$

$$90 =$$

$$LCM =$$

Example 2. Find the LCM of 12, 30, and 70.

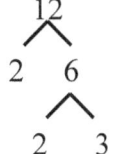

Step 1. Find the prime factorization of each number.

$$12 = 2 \cdot 2 \cdot 3$$

$$30 = 2 \cdot 3 \cdot 5$$

$$70 = 2 \cdot 5 \cdot 7$$

$$LCM = 2 \cdot 2 \cdot 3 \cdot 5 \cdot 7$$

$$= \boxed{420}$$

Step 2. Write "LCM=" below the prime factorization of each number.

Consider each prime number. Write it the greatest number of times it occurs in any one factorization.

2: it occurs twice.

3: it occurs once.

5: it occurs once.

7: it occurs once.

Now multiply. This is the LCM.

Your Turn Problem #2

Find the LCM of 24, 150 and 240.

$$24 =$$

$$150 =$$

$$240 =$$

$$LCM =$$

Example 3. Find the LCM of 7 and 15.

$$7 \qquad\qquad 15$$

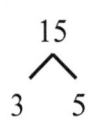

prime $\qquad\qquad$ 3 $\quad$ 5

$7 = 7$

$15 = 3 \cdot 5$

$LCM = 3 \cdot 5 \cdot 7$

$= \boxed{105}$

Step 1. Find the prime factorization of each number.

Step 2. Write "LCM=" below the prime factorization of each number.

Consider each prime number. Write it the greatest number of times it occurs in any one factorization.

3: it occurs once.

5: it occurs once.

7: it occurs once.

Now multiply. This is the LCM.

Your Turn Problem #3

Find the LCM of 11 and 25.

$\qquad$ 11 =

$\qquad$ 25 =

$\qquad$ LCM =

Example 4. Find the LCM of 39, 45, and 65.

$$39 = 3 \cdot 13$$
$$45 = 3 \cdot 3 \cdot 5$$
$$65 = 5 \cdot 13$$
$$LCM = 3 \cdot 3 \cdot 5 \cdot 13$$
$$= \boxed{585}$$

Step 1. Find the prime factorization of each number.

Step 2. Write "LCM=" below the prime factorization of each number.

Consider each prime number. Write it the greatest number of times it occurs in any one factorization.

3: it occurs twice.
5: it occurs once.
13: it occurs once.
Now multiply. This is the LCM.

Your Turn Problem #4

Find the LCM of 10, 46 and 115.

$$10 =$$

$$46 =$$

$$115 =$$

$$LCM =$$

3.9 Homework: The LCM (Least Common Multiple)

Find the LCM for the following.

1. 9, 15

2. 18, 24, 27

3. 35, 45

4. 15, 19, 24

5. 18, 24

6. 8, 20

7. 16, 20

8. 15, 24

3.9 Homework: The LCM (Least Common Multiple) cont.

Find the LCM for the following.

9. 35, 75

10. 12, 18, 24

11. 24, 36

12. 84, 90

13. 35, 45, 63

14. 30, 48

3.9 Homework: The LCM (Least Common Multiple) cont.

Find the LCM for the following.

15. 9,12, and 18

16. 15, 20 and 6

17. 12, 24, and 15

18. 27, 24, and 36

19. 15, 33, and 55

20. 12, 18, and 24

Practice Test 3

Solve the following applications.

1. A bookstore ordered 420 copies of a textbook at a cost of $9,660. What was the cost to the store of each textbook?

2. Diego agreed to buy a new car paying $1200 down and $290 a month for 5 years. What is the total cost of the car?

3. A computer printer can print 43 pages per minute. How many pages can be printed in 1 hour?

4. There are 415 students who are taking a field trip. If each bus can hold 40 students, how many buses will be needed for the field trip?

5. A truck holds 12 gallons of gas. If the truck gets 19 miles per gallon, how far can the truck go on a full tank of gas?

Practice Test 3 cont.

Solve the following applications.

6. Alisa earns $940 per week. How much money would she earn in 52 weeks?

7. Larry has 5 math exams with scores of 85, 72, 63, 91, and 94. Find the average of all five exams.

8. Find the perimeter of the rectangle if the width is 7 ft and the length is 12 ft.

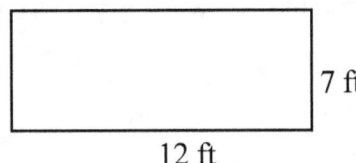

7 ft

12 ft

9. Find the area of the rectangle if the width is 3 m and the length is 16 m.

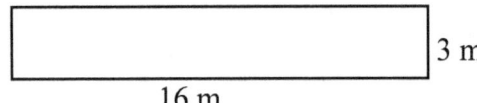

3 m

16 m

Practice Test 3 cont.

Solve the following applications.

10. Find the area of the triangle if the base is 40 ft and the height is 15 ft.

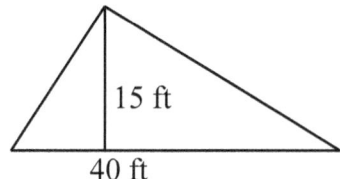

15 ft

40 ft

11. Find the volume of the box:

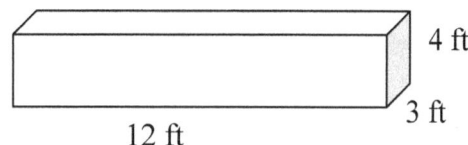

4 ft

3 ft

12 ft

12. Determine if 59 is a prime number. State you reasoning for "if it is or is not a prime number".

13. Determine if 231 is a prime number. State you reasoning for "if it is or is not a prime number".

Find the prime factorization for each number.
14. 48   **15.** 240   **16.** 525

Practice Test 3 cont.

17. List all the factors of 30.

18. List all the factors of 45.

Find the GCF for the following groups of numbers.

19. 24 and 36

20. 42 and 70

Find the LCM for the following groups of numbers.

21. 28 and 35

22. 15 and 24

23. 7 and 18

24. 20 and 45

4.1 Introduction to Fractions

Definition: Fraction (or a Rational Number)

A rational number is a number that can be written in the fraction form $\frac{a}{b}$ where a is a whole number and b is a number that is not zero. The top number is called the numerator and the bottom number is called the denominator.

$$\frac{a}{b} \begin{array}{l} \leftarrow \text{numerator} \\ \leftarrow \text{denominator} \end{array} \qquad \text{Examples:: } \quad \frac{3}{4}, \frac{5}{7}, \frac{87}{100}$$

In this course, when we say fractions, we really mean rational numbers. In later courses, we will expand upon the definition of a rational number. Rational numbers (fractions) may contain integers (positive and negative numbers). We will also cover irrational numbers. An example of irrational numbers are $\sqrt{5}$ (read square root of 5), and π (read pi, $\pi \approx 3.14\ldots\ldots$). Since these are going to be covered in a later course, we simply won't worry about those now.

How to say it and write it.

A fraction can always be read as: "numerator over denominator."

Examples: $\frac{12}{55}$ is read as 12 over 55.

$\frac{42}{97}$ is read 42 over 97.

$\frac{3}{8}$ is read 3 over 8.

A more grammatically correct form would include the use of hyphens with the numbers written in words. A hyphen is used to separate the numerator and the denominator. However, if either the numerator or denominator is already hyphenated, then a hyphen is not used to separate the numerator and denominator. Also, the following suffixes are used on words representing the denominators: th, st, nd, and rd. The denominators will be changed to an ordinal form of the number.

Examples: "Third" is the ordinal form of three. "Eighth" is the ordinal form of eight.

Examples: a) $\dfrac{3}{16}$ is written or verbalized as three–sixteenths.

b) $\dfrac{2}{3}$ is written or verbalized as two–thirds.

c) $\dfrac{5}{21}$ is written or verbalized as five twenty–firsts.

d) $\dfrac{9}{28}$ is written or verbalized as nine twenty–eighths.

Notice the hyphen separating the numerator and denominator in a) and b).

There is no hyphen separating the numerator and denominator in c) and d) because the denominator already contains a hyphen.

Fractions can be used in two basic ways:

1. To indicate equal parts of a whole.

2. To indicate division. (The numerator is to be divided by the denominator.)

Fractions used to indicate equal parts of a whole.

For example: 3/8 implies that the whole is divided into eight equal parts. The numerator indicates a portion of the equal parts. That is, 3/8 would indicate 3 of 8 equal parts.

$\dfrac{3}{8}$ (3 out of 8 are shaded)

Therefore the top number of the fraction is representative of the portion to be described or expressed and the bottom number is representative of the number of divisions, sections, or portions in which the whole has been divided.

For example: $\dfrac{3}{8} = \dfrac{\text{portion being expressed}}{\text{number of equal sections whole has been partioned}}$

Example 1. What fraction indicates the shaded part of the diagram?

Answer: Since there are 12 equal squares and 5 are shaded, the shaded part is $\dfrac{5}{12}$.

Your Turn Problem #1

Shade $\dfrac{7}{10}$ of the rectangle.

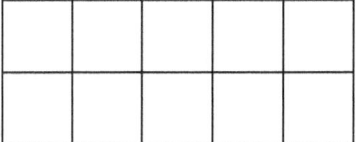

Writing a fraction from a sentence.

The top number (numerator) represents the portion described. The bottom number (denominator) can be thought of as the total.

Example 2: Write the desired fraction.

a) Five out of the seven boxes are shaded. What fraction of the boxes are shaded?

    Answer $\dfrac{5}{7}$ of the boxes are shaded.

b) Three out of twenty residents have a pool. What fraction of the residents have a pool?

    Answer: $\dfrac{3}{20}$ of the residents have a pool.

Example 2 cont: Write the desired fraction.

    c)   Five out seven boxes are shaded. What fraction of the boxes are not shaded?

       Answer: $\dfrac{2}{7}$ of the boxes are not shaded. (Since 5 are shaded, 2 are not shaded.)

    d)   Three out of eight voters voted in favor of a certain proposition. What fraction of the voters voted against the proposition?

       Answer: $\dfrac{5}{8}$ of the voters voted against the proposition.

Your Turn Problem #2

Write the desired fraction.

a) Five out of twelve students have cars. What fraction of the students have cars?

Answer:_____

b) Nine out of sixteen students have internet at home. What fraction of the students do not have internet at home?

Answer:_____

c) Marissa missed three questions on a twenty question test. What fraction of the questions did she get correct?

Answer:_____

A fraction can be used to indicate division.

Since $\dfrac{a}{b}$ can be described as $a \div b$, b cannot be zero since division by zero is not possible. It is

undefined. $\left(\dfrac{a}{b} \rightarrow b\overline{)a} \quad \text{or} \quad \text{denominator}\overline{)\text{numerator}} \right)$

Example: $\dfrac{12}{3}$ is equivalent to $12 \div 3$ which equals 4.

Example 3a. Simplify: $\dfrac{28}{4}$

Answer: $\dfrac{28}{4}$ is equivalent to $28 \div 4$ which equals 7.

Example 3b. Simplify: $\dfrac{51}{3}$

Answer: $\dfrac{51}{3}$ is equivalent to $51 \div 3$ which equals 17.

Example 3c. Simplify: $\dfrac{125}{5}$

Answer: $\dfrac{125}{5}$ is equivalent to $125 \div 5$ which equals 25.

Your Turn Problem #3

Simplify the following.

a) $\dfrac{145}{5}$
b) $\dfrac{87}{3}$
c) $\dfrac{91}{7}$

Answer: _____ Answer: _____ Answer: _____

Fractions Equaling "1"

Any fraction with a non-zero denominator that is equal to its numerator is equal to one. If portion is the same as the number of divisions, then the amount described is the whole.

Example 4a. Simplify: $\dfrac{12}{12}$

Answer: $\dfrac{12}{12}$ is equivalent to $12 \div 12$ which equals 1. Any number divided by itself equals 1.

Example 4b. Simplify: $\dfrac{121}{121}$

Answer: $\dfrac{121}{121}$ is equivalent to $121 \div 121$ which equals 1. Any number divided by itself equals 1.

Your Turn Problem #4

Simplify the following.

a) $\dfrac{7}{7}$　　　　　　　　　b) $\dfrac{39}{39}$　　　　　　　　　c) $\dfrac{156}{156}$

Answer: _____　　　　　Answer: _____　　　　　Answer: _____

Fractions where the denominator equals "1".

Any fraction with a denominator that is equal to 1 is equal to the numerator. Any number divided by 1 is itself.

Example 5a. Simplify: $\dfrac{5}{1}$

Answer: $\dfrac{5}{1}$ is equivalent to $5 \div 1$ which equals 5. Any number divided by 1 equals itself.

Example 5b. Simplify: $\dfrac{23}{1}$

Answer: $\dfrac{23}{1}$ is equivalent to $23 \div 1$ which equals 23. Any number divided by 1 equals itself.

Your Turn Problem #5

Simplify the following.

a) $\dfrac{8}{1}$

b) $\dfrac{1}{1}$

c) $\dfrac{356}{1}$

Answer: _____

Answer: _____

Answer: _____

Fractions where the numerator equals "1".

Any fraction with a numerator equal to 1 can not be simplified (denominator is not equal zero).

The fraction $\dfrac{1}{3}$ means one out of three. $\dfrac{1}{3}$ does not simplify. Many students make the mistake

of writing $\dfrac{1}{3} = 3$. This is incorrect.

Example 6a. Simplify: $\dfrac{1}{8}$

Answer: $\dfrac{1}{8}$

Example 6b. Simplify: $\dfrac{1}{16}$

Answer: $\dfrac{1}{16}$

Your Turn Problem #6

Simplify the following.

a) $\dfrac{1}{12}$

b) $\dfrac{1}{4}$

c) $\dfrac{1}{153}$

Answer: _____

Answer: _____

Answer: _____

Fractions Equaling "0"

Any fraction with zero in the numerator and a denominator that is not zero is equal to zero. If nothing is shaded, then the amount described is zero.

Example 7a. Simplify: $\dfrac{0}{5}$

Answer: $\dfrac{0}{5}$ is equivalent to $0 \div 5$ which equals 0.

Zero divided by any number (except 0) equals 0.

Example 7b. Simplify: $\dfrac{0}{18}$

Answer: $\dfrac{0}{18}$ is equivalent to $0 \div 18$ which equals 0.

Zero divided by any number (except 0) equals 0.

Your Turn Problem #7

Simplify the following.

a) $\dfrac{0}{10}$

b) $\dfrac{0}{1}$

c) $\dfrac{0}{225}$

Answer: _____

Answer: _____

Answer: _____

Fractions where the Denominator Equals "0"

An object being portioned into zero parts certainly does not make sense. We call a fraction "undefined" when the denominator is equal to zero but the numerator does not equal zero. The word "undefined" must be used to describe such a fraction. Do not use any other words or notation. This is also consistent with our rules of division. We have to be able to write a related multiplication sentence that is true.

For example: $\dfrac{12}{4} = 3 \rightarrow 12 \div 4 = 3$. The related multiplication sentence is $3 \cdot 4 = 12$.

$\dfrac{12}{0} = ?$ Many students want to write an answer of zero. This is <u>not correct</u>. The related multiplication sentence is false.

$\dfrac{12}{0} = 0 \rightarrow 12 \div 0 = 0$ The related multiplication sentence is $0 \cdot 0 = 12$, which is false. Therefore, we can not write 0 as an answer. The correct answer would be "undefined."

Example 8a. Simplify: $\dfrac{9}{0}$

Answer: $\dfrac{9}{0}$ is undefined.

Example 8b. Simplify: $\dfrac{15}{0}$

Answer: $\dfrac{15}{0}$ is undefined.

Your Turn Problem #8

Simplify the following if possible.

a) $\dfrac{13}{0}$

b) $\dfrac{1}{0}$

c) $\dfrac{153}{0}$

Answer: _____

Answer: _____

Answer: _____

Note: If the numerator and denominator both equal zero, then the fraction is called **indeterminate.** $\dfrac{0}{0}$ is called indeterminate. This concept will be studied further in a calculus course.

Example 9. Simplify: $\dfrac{0}{0}$

Answer: $\dfrac{0}{0}$ is indeterminate.

Your Turn Problem #9

Simplify the following if possible: $\dfrac{0}{0}$

Answer: _____

Definition: Proper Fraction

A proper fraction is a fraction whose numerator is less than the denominator.

Examples: $\dfrac{1}{2}, \dfrac{3}{5}, \dfrac{7}{25}$

Definition: Improper Fraction

An improper fraction is a fraction whose numerator is greater than *or equal* to the denominator.

Examples: $\dfrac{8}{5}, \dfrac{8}{8}, \dfrac{277}{25}$

Definition: Mixed Number

A mixed number is a number with a whole number and a proper fraction.

Example: $4\dfrac{1}{2}, 3\dfrac{3}{5}, 90\dfrac{7}{25}$

Converting an Improper Fraction to a Mixed Number

To change an improper fraction to a mixed number:

Divide the denominator into the numerator. Write the remainder as a fraction by placing the remainder in the numerator and the original denominator as the denominator of the fraction part of the mixed number.

$$\text{denominator} \overline{)\text{numerator}} \Rightarrow \text{quotient} \frac{\text{remainder}}{\text{divisor}}$$

Example 10a: Convert $\dfrac{9}{4}$ to a mixed number.

Answer: a) $\begin{array}{r} 2 \\ 4\overline{)9} \\ -8 \\ \hline 1 \end{array}$ $= \boxed{2\dfrac{1}{4}}$

Example 10b: Convert $\dfrac{79}{8}$ to a mixed number.

Answer: a) $\begin{array}{r} 9 \\ 8\overline{)79} \\ -72 \\ \hline 7 \end{array}$ $= \boxed{9\dfrac{7}{8}}$

Your Turn Problem #10

Convert the following to mixed numbers.

a) $\dfrac{18}{7}$ 　　　　　　 b) $\dfrac{35}{4}$ 　　　　　　 c) $\dfrac{57}{11}$

Converting a Mixed Number to an Improper Fraction

To change a mixed number into an improper fraction, multiply the denominator by the whole number and then add that product to the numerator. This sum becomes the numerator of the improper fraction and the denominator of the improper fraction is the same as the denominator from the mixed number.

Example 11a. Convert $5\dfrac{2}{3}$ to an improper fraction.

$5\dfrac{2}{3} \Rightarrow$ Step 1   Multiply the whole number and denominator. $5 \cdot 3 = 15$.

Then add the numerator to the 15. $15 + 2 = 17$. This number is the numerator of the improper fraction.

Step 2.   The denominator is unchanged. It remains a 3.

Answer: $\boxed{\dfrac{17}{3}}$

Example 11b. Convert $16\dfrac{3}{5}$ to an improper fraction.

$16\dfrac{3}{5} \Rightarrow$ Step 1.   Multiply the whole number and denominator. $16 \cdot 5 = 80$.

Then add the numerator to the 80. $80 + 3 = 83$. This number is the numerator of the improper fraction.

Step 2.   The denominator is unchanged. It remains a 5.

Answer: $\boxed{\dfrac{83}{5}}$

Your Turn Problem #11

Convert the following to improper fractions.

a) $9\dfrac{1}{2} =$ b) $12\dfrac{3}{4} =$ c) $17\dfrac{2}{3} =$

4.1 Homework: Introduction to Fractions

1. Identify the numerator and denominator of the fraction: $\dfrac{6}{13}$

What fraction names the shaded part of each of the following figures?

2.

3.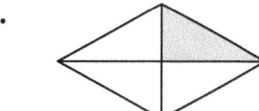

4. You missed 5 questions on a 12 question test. What fraction names the part you got correct?

5. A used car dealer sold 11 of the 17 cars in stock. What fraction names the portion sold?

6. A used car dealer sold 11 of the 17 cars in stock. What fraction names the portion not sold?

7. If a pizza has 12 slices and 5 slices are "missing", what fraction is missing?

8. In a class of 35 math students, only 19 passed the first exam. What fraction of the class failed the first exam?

9. A student got two A's, four B's and three C's. What fraction of her test scores were A's?

10. Out of a dozen roses, 3 are red, 5 are pink, and 4 are white. What fraction of the roses are pink?

4.1 Homework: Introduction to Fractions

11. In a shipment of 70 alternators, 9 were found to be defective. What fraction of the alternators were defective?

12. Scott made 16 out of 21 free throws. What fraction of the free throws did he miss?

13. $\frac{7}{9}$ is an example of a _____ .

14. $\frac{12}{5}$ is an example of an _____ .

15. $7\frac{1}{8}$ is an example of a _____ .

16. A fraction where the numerator is equal to the denominator is always equal to _____ as long the denominator does not equal zero.

17. A fraction whose denominator is equal to zero and whose numerator is not zero is called

_____ .

Identify each number as a proper fraction, an improper fraction or a mixed number.

18. $\frac{3}{5}$:

19. $3\frac{2}{5}$:

20. $\frac{11}{4}$:

21. $\frac{6}{6}$:

22. $\frac{48}{7}$:

23. $\frac{13}{17}$:

4.1 Homework: Introduction to Fractions cont.

Simplify the following if possible. (i.e., divide the denominator into the numerator.)

24. $\dfrac{24}{6}$

25. $\dfrac{45}{1}$

26. $\dfrac{7}{7}$

27. $\dfrac{95}{5}$

28. $\dfrac{161}{7}$

29. $\dfrac{696}{12}$

30. $\dfrac{35}{1}$

31. $\dfrac{38}{2}$

32. $\dfrac{88}{8}$

33. $\dfrac{17}{0}$

34. $\dfrac{0}{2}$

35. $\dfrac{33}{33}$

36. $\dfrac{51}{3}$

37. $\dfrac{29}{0}$

4.1 Homework: Introduction to Fractions cont.

Change the following improper fractions to mixed numbers.

38. $\dfrac{7}{5}$

39. $\dfrac{22}{9}$

40. $\dfrac{24}{5}$

41. $\dfrac{43}{7}$

42. $\dfrac{38}{9}$

43. $\dfrac{57}{11}$

44. $\dfrac{73}{6}$

45. $\dfrac{40}{17}$

46. $\dfrac{113}{25}$

47. $\dfrac{95}{24}$

48. $\dfrac{77}{5}$

49. $\dfrac{96}{11}$

4.1 Homework: Introduction to Fractions cont.

Change the following mixed numbers to improper fractions.

50. $4\dfrac{2}{5}$

51. $6\dfrac{2}{7}$

52. $9\dfrac{1}{2}$

53. $122\dfrac{1}{3}$

54. $17\dfrac{4}{5}$

55. $1\dfrac{1}{2}$

56. $33\dfrac{1}{3}$

57. $22\dfrac{2}{9}$

58. $16\dfrac{2}{3}$

59. $81\dfrac{1}{2}$

4.2 Simplifying Fractions

Equivalent Fractions

Fractions that have the same value are equivalent fractions.

For example: $\dfrac{6}{12} = \dfrac{50}{100} = \dfrac{1}{2}$

A fraction written in *lowest terms* is when the fraction is expressed using the lowest numerator and denominator possible. For instance, in the example from the *equivalent fractions* used above, 1/2 represents that fraction written in lowest terms because it is impossible to find an equivalent fraction that represents that value using a smaller numerator or denominator.

Reducing or Simplifying

Reducing is the process of converting a fraction to an equivalent fraction in lowest terms. It is done by dividing the numerator and denominator by the <u>same</u> factor or factors until there are no further factors that can be evenly divided into each.

Reducing Fractions to Lowest Terms by Dividing by Common Factors

Reducing is done by dividing the numerator and denominator by the largest factor that divides into both numbers. The largest number that divides evenly into a set of numbers is called the GCF, (greatest common factor). Actually, it is not absolutely necessary to use the GCF. It will just take more steps if we don't use the GCF.

Example 1a. Simplify $\dfrac{36}{42}$

    1.   What is the GCF of 36 and 42? Answer: 6

    2.   Divide both numerator and denominator by 6.

$$\dfrac{36 \div 6}{42 \div 6}$$

Answer: $\boxed{\dfrac{6}{7}}$

Recall: Finding the GCF

Step 1. Find the prime factorization of each number.

$$36 = 2 \cdot 2 \cdot 3 \cdot 3$$
$$42 = 2 \cdot 3 \cdot 7$$

Step 2. Circle any prime that appears in both rows.

Step 3. Multiply the common primes together from one row.

The GCF $= 2 \cdot 3 = 6$

Although we will now show an alternative to using the GCF, the concept of finding the GCF should be comprehended. The GCF will also be used in algebra when factoring polynomials is introduced. Don't worry, you will get to factor polynomials soon enough. ☺

Alternative solution: Not using the GCF

Find *any* number that divides evenly into both numerator and denominator.

Since both numbers are even, 2 divides evenly into both numbers.

$$\frac{36 \div 2}{42 \div 2} = \frac{18}{21}$$

Is there a number that divides evenly into 18 and 21? Yes, it is 3. Divide both by 3.

$$\frac{18 \div 3}{21 \div 3} = \frac{6}{7}$$

We still got the same answer, it just took one more step but that is not a problem.

Example 1b. Simplify $\dfrac{105}{165}$

1. What is the GCF of 105 and 165? Answer: 15

2. Divide both numerator and denominator by 15.

$$\frac{105 \div 15}{165 \div 15}$$

Answer: $\boxed{\dfrac{7}{11}}$

The use of the divisibility rules here can also be helpful when trying to find a number that divides evenly into both numbers. Recall the following divisibility rules:

2: 2 divides evenly into numbers that end with an even digit.

3: If the sum of the digits is divisible by 3, then 3 divides evenly into the number.

5: If a number ends in a 0 or a 5, then 5 divides evenly into the number.

7: No divisibility test given. Easier to just divide to see if the number is divisible by 7.

Example 1b again. Simplify $\dfrac{105}{165}$

Using our divisibility rules, since both numbers end in a 0 or a 5, we know 5 will divide evenly into both numbers.

$$\frac{105 \div 5}{165 \div 5} = \frac{21}{33}$$

Now we want to reduce $\dfrac{21}{33}$. Start with the prime numbers to find if both numbers are divisible by the same number. 3 divides evenly into both numbers.

$$\frac{21 \div 3}{33 \div 3} = \boxed{\frac{7}{11}}$$

Your Turn Problem #1

Simplify the following:

a) $\dfrac{45}{60}$

b) $\dfrac{132}{180}$

Prime Factorization Method of Reducing

This method can be more efficient when the fraction contain large numbers. This method requires writing the prime factorization of both numerator and denominator. Since any number divided by itself equals one, we will line out any like factors that are contained in both numerator and denominator.

Procedure for Reducing using Prime Factorization

Step 1: Find the prime factorization of the numerator and denominator.

Step 2: Rewrite the fraction using the prime factorizations.

Step 3: Line out, one for one, like factors that are contained both in the numerator and denominator. (Any number divided by itself equals 1.)

Step 4: Multiply the remaining factors that are not lined-out together.

Example 2a. Simplify $\dfrac{102}{136}$

1. Find the prime factorization of each number.

$$102 = 2 \cdot 3 \cdot 17$$
$$136 = 2 \cdot 2 \cdot 2 \cdot 17$$

$$\begin{array}{r} 17 \\ 3\overline{)51} \\ 2\overline{)102} \end{array} \qquad \begin{array}{r} 17 \\ 2\overline{)34} \\ 2\overline{)68} \\ 2\overline{)136} \end{array}$$

2. Rewrite the fraction using the prime factorizations. $\dfrac{102}{136} = \dfrac{2 \cdot 3 \cdot 17}{2 \cdot 2 \cdot 2 \cdot 17}$

3. Line out like factors. (Any number over itself equals 1.) $\dfrac{\cancel{2} \cdot 3 \cdot \cancel{17}}{\cancel{2} \cdot 2 \cdot 2 \cdot \cancel{17}}$

4. Multiply the remaining factors.

Answer: $\boxed{\dfrac{3}{4}}$

Note: This is a <u>very methodical procedure for reducing fractions</u>. All it requires is to do the prime factorization tree (or the division method). When writing out the prime factorization tree, keep in mind the divisibility rules. The divisibility rules will help with the prime factorization tree.

Example 2b. Simplify $\dfrac{190}{228}$

    1.   Find the prime factorization of each number.

$$190 = 2 \cdot 5 \cdot 19$$
$$228 = 2 \cdot 2 \cdot 3 \cdot 19$$

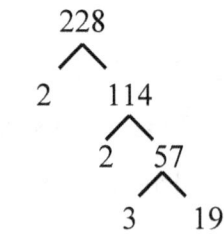

    2.   Rewrite the fraction using the prime factorizations. $\dfrac{190}{228} = \dfrac{2 \cdot 5 \cdot 19}{2 \cdot 2 \cdot 3 \cdot 19}$

    3.   Line out like factors. (Any number over itself equals 1.) $\dfrac{\cancel{2} \cdot 5 \cdot \cancel{19}}{\cancel{2} \cdot 2 \cdot 3 \cdot \cancel{19}}$

    4.   Multiply the remaining factors.

Answer: $\boxed{\dfrac{5}{6}}$

Your Turn Problem #2

Simplify the following:

a) $\dfrac{95}{114}$

b) $\dfrac{98}{105}$

What if all the numbers in the numerator or denominator all line out?

When we line out, the factor becomes 1 (any number divided by itself equals 1). If a numerator or denominator is completely lined out, its result is the product of 1's which is equal to "1.

Example 3a. Simplify $\dfrac{5}{60}$

1. Find the prime factorization of each number.

$5 = 5$ (prime)

$60 = 2 \cdot 2 \cdot 3 \cdot 5$

$$\begin{array}{r} 5 \\ 3\overline{)15} \\ 2\overline{)30} \\ 2\overline{)60} \end{array}$$

2. Rewrite the fraction using the prime factorizations. $\quad \dfrac{5}{60} = \dfrac{5}{2 \cdot 2 \cdot 3 \cdot 5}$

3. Line out like factors (any number over itself equals 1) $\quad \dfrac{\cancel{5}}{2 \cdot 2 \cdot 3 \cdot \cancel{5}}$

4. Multiply the remaining factors. Remember, if all of the numbers line out in either numerator or denominator, the result in that position is a 1.

Answer: $\boxed{\dfrac{1}{12}}$

Example 3b. Simplify $\dfrac{28}{14}$

1. Find the prime factorization of each number.

$28 = 2 \cdot 2 \cdot 7$

$14 = 2 \cdot 7$

$$\begin{array}{r} 7 \\ 2\overline{)14} \\ 2\overline{)28} \end{array} \qquad \begin{array}{r} 7 \\ 2\overline{)14} \end{array}$$

2. Rewrite the fraction using the prime factorizations. $\quad \dfrac{28}{14} = \dfrac{2 \cdot 2 \cdot 7}{2 \cdot 7}$

3. Line out like factors (any number over itself equals 1) $\quad \dfrac{2 \cdot \cancel{2} \cdot \cancel{7}}{\cancel{2} \cdot \cancel{7}}$

4. Multiply the remaining factors. Remember, if all of the numbers line out in either numerator or denominator, the result in that position is a 1.

Answer: $\dfrac{2}{1} = \boxed{2}$ (Any number divided by 1 is itself.)

If all of the factors in the denominator are lined out, then the denominator is equal to 1. Since any number divided by 1 will be the number itself, the result will be the product of the factors in the numerator. In other words, the answer is a whole number.

If all of the factors in the numerator are lined out, then the numerator is equal to 1. Since 1 over any number (except zero) can not be simplified, the result will be the 1 over the product of the factors in the denominator.

Your Turn Problem #3

Simplify the following:

a) $\dfrac{15}{90}$

b) $\dfrac{42}{7}$

c) $\dfrac{39}{156}$

Reducing Improper Fractions

Improper fractions are fractions where the numerator is greater than the denominator.

For example, $\dfrac{8}{5}$ is an improper fraction. It can be converted to a mixed number, but it can not be reduced. For a fraction to reduce, a number has to divide evenly into both numerator and denominator. 8 and 5 do not have a common factor. If we wish to convert $\dfrac{8}{5}$ to mixed number, the answer is $1\dfrac{3}{5}$. Please realize that converting an improper fraction to a mixed number is not reducing a fraction.

If an improper fraction can be reduced, then it must be reduced. Actually, any answer that can be simplified must be simplified even if the directions do not specify.

An improper fraction should be converted to a mixed number if the directions state so.

An improper fraction should also be converted to a mixed number if the answer is an answer to an application. For example, we would not answer "It takes $\dfrac{5}{2}$ hours to drive to Camarillo."

It would make more sense to answer "It takes $2\dfrac{1}{2}$ hours to drive to Camarillo."

Procedure for Reducing and Converting an Improper Fraction to a Mixed Number.

1. Reduce the fraction.
2. Convert the improper fraction to a mixed number.

    or

1. Convert the improper fraction to a mixed number.
2. Reduce the fraction.

It does not matter which step is done first.

Example 4. Simplify $\dfrac{24}{10}$

In this example, we will reduce first, then convert to a mixed number.

1. Find the prime factorization of each number.

$24 = 2 \cdot 2 \cdot 2 \cdot 3$

$10 = 2 \cdot 5$

$$\begin{array}{r} 3 \\ 2\overline{)6} \\ 2\overline{)12} \\ 2\overline{)24} \end{array} \qquad \begin{array}{r} 5 \\ 2\overline{)10} \end{array}$$

2. Rewrite the fraction using the prime factorizations. $\dfrac{24}{10} = \dfrac{2 \cdot 2 \cdot 2 \cdot 3}{2 \cdot 5}$

3. Line out like factors. (Any number over itself equals 1.) $\dfrac{\cancel{2} \cdot 2 \cdot 2 \cdot 3}{\cancel{2} \cdot 5}$

4. Multiply the remaining factors. $\dfrac{12}{5}$

5. Convert the improper fraction to a mixed number. (denominator $\overline{)\,\text{numerator}}$)

Answer: $\boxed{2\dfrac{2}{5}}$

Your Turn Problem #4

Simplify $\dfrac{90}{25}$

Example 5. Simplify $\dfrac{56}{20}$

In this example, we will convert to a mixed number first, then reduce the fraction.

1. Convert the improper fraction to a mixed number. ($\overline{\text{denominator}\,)\,\text{numerator}}$)

$$\frac{56}{20} \;\rightarrow\; \begin{array}{r} 2 \\ 20\overline{)56} \\ -40 \\ \hline 16 \end{array} \;\rightarrow\; 2\frac{16}{20}$$

2. Now, we need to reduce the fractional part of the mixed number.

   Find the prime factorization of each number.

   $16 = 2\cdot2\cdot2\cdot2$
   $20 = 2\cdot2\cdot5$

$$\begin{array}{r} 2 \\ 2\overline{)4} \\ 2\,\overline{)8} \\ 2\,\overline{)16} \end{array} \qquad \begin{array}{r} 5 \\ 2\overline{)10} \\ 2\,\overline{)20} \end{array}$$

3. Rewrite the fraction using the prime factorizations. $\dfrac{16}{20} = \dfrac{2\cdot2\cdot2\cdot2}{2\cdot2\cdot5}$

4. Line out like factors. (Any number over itself equals 1.) $\dfrac{\cancel{2}\cdot\cancel{2}\cdot2\cdot2}{\cancel{2}\cdot\cancel{2}\cdot5}$

5. Multiply the remaining factors. $\dfrac{4}{5}$

6. Write the answer with the whole number in front of the reduced fraction.

Answer: $\boxed{2\dfrac{4}{5}}$

Your Turn Problem #5

Simplify $\dfrac{135}{40}$

Recall: Writing a fraction from a sentence.

The top number (numerator) represents the portion described. The bottom number (denominator) can be thought of as the total.

Example 6. 24 out of 30 students passed the first test. What fraction of the students passed the first test?

Solution: Numerator represents the portion described, 24. The denominator represents the total, which is 30.

$$\frac{24}{30} = \frac{\cancel{2} \cdot 2 \cdot 2 \cdot \cancel{3}}{\cancel{2} \cdot \cancel{3} \cdot 5} = \frac{4}{5}$$

Answer: $\boxed{\dfrac{4}{5}}$ of the students passed the first test.

Your Turn Problem #6

50 out of 65 employees work the day shift. What fraction of the employees do not work the day shift?

4.2 Homework: Simplifying Fractions

Write each fraction is simplest form if possible. If it is already in lowest terms, simply rewrite the fraction.

1. $\dfrac{18}{24}$

2. $\dfrac{33}{45}$

3. $\dfrac{48}{60}$

4. $\dfrac{38}{57}$

5. $\dfrac{9}{25}$

6. $\dfrac{44}{77}$

7. $\dfrac{54}{72}$

8. $\dfrac{18}{36}$

4.2 Homework: Simplifying Fractions cont

Write each fraction is simplest form if possible. If it is already in lowest terms, simply rewrite the fraction.

9. $\dfrac{15}{22}$

10. $\dfrac{120}{720}$

11. $\dfrac{0}{12}$

12. $\dfrac{55}{1}$

13. $\dfrac{75}{125}$

14. $\dfrac{121}{165}$

15. $\dfrac{66}{135}$

16. $\dfrac{57}{76}$

4.2 Homework: Simplifying Fractions cont

Write each fraction is simplest form if possible. If it is already in lowest terms, simply rewrite the fraction.

17. $\dfrac{46}{69}$

18. $\dfrac{74}{185}$

19. $\dfrac{22}{22}$

20. $\dfrac{14}{0}$

21. $\dfrac{124}{155}$

22. $\dfrac{144}{216}$

23. $\dfrac{58}{74}$

24. $\dfrac{57}{75}$

4.2 Homework: Simplifying Fractions cont

Write each fraction is simplest form if possible. If it is already in lowest terms, simply rewrite the fraction.

25. $\dfrac{150}{210}$

26. $\dfrac{203}{232}$

27. $\dfrac{49}{91}$

28. $\dfrac{57}{95}$

29. $\dfrac{102}{170}$

30. $\dfrac{434}{496}$

31. $\dfrac{38}{57}$

32. $\dfrac{29}{129}$

4.2 Homework: Simplifying Fractions cont

Write each fraction is simplest form if possible. If it is already in lowest terms, simply rewrite the fraction.

33. $\dfrac{1400}{1600}$

34. $\dfrac{357}{408}$

35. $\dfrac{24}{48}$

36. $\dfrac{17}{51}$

Simplify each improper fraction. Leave answer as a mixed number.

37. $\dfrac{22}{10}$

38. $\dfrac{40}{15}$

39. $\dfrac{30}{8}$

40. $\dfrac{133}{38}$

4.2 Homework: Simplifying Fractions cont

Simplify each improper fraction. Leave answer as a mixed number.

41. $\dfrac{46}{22}$

42. $\dfrac{80}{15}$

43. $\dfrac{100}{12}$

44. $\dfrac{150}{45}$

45. $\dfrac{99}{55}$

46. $\dfrac{96}{84}$

47. $\dfrac{140}{25}$

48. $\dfrac{100}{34}$

4.2 Homework: Simplifying Fractions cont

49. In a shipment of 80 digital cameras, 12 were found to be defective. What fraction of the cameras were defective?

50. On a 40-question test, Elaine missed 8 questions. What fraction of the questions did she get correct?

51. A study showed that 110 students out of 550 students in an elementary school were left-handed. What fraction of the students were left-handed?

52. A girls' softball team played 48 games. If they lost only 6 games, what fraction of the games did the team win?

53. Scott made 24 out of 30 free throws. What fraction of the free throws did he make?

4.3 Reading a Ruler

A standard tape measure (or ruler) is divided into feet and inches. Each foot is divided into 12 inches and each inch is labeled on the ruler (1, 2, 3, 4, …). Each inch is equally partitioned. Rulers are portioned into fourths, eighths, or sixteenths. The ruler below is partitioned into fourths. After the zero, each line (called a tick mark) is one-fourth of an inch.

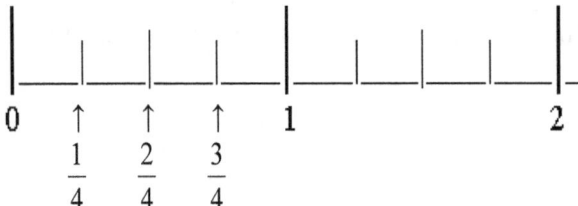

The second tick mark (after zero) is $\frac{2}{4}$ of an inch, which can be reduced to $\frac{1}{2}$ in.

Notice that the $\frac{1}{2}$ inch mark (also called half-inch mark) is longer than the $\frac{1}{4}$ in and $\frac{3}{4}$ in. The next tick mark is the longest line at the 1" (" means inch). This is consistent with the ordering of the fraction; $\frac{1}{4}$, $\frac{2}{4}$, $\frac{3}{4}$, and $\frac{4}{4}$. Any number divided by itself is 1. The first tick mark after 1" is $1\frac{1}{4}$ in.

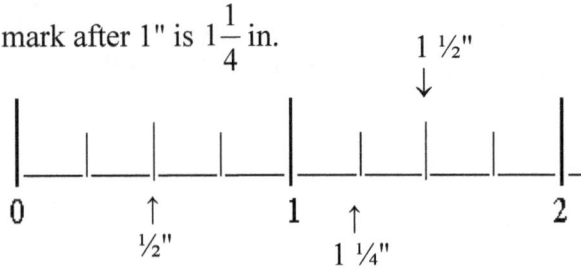

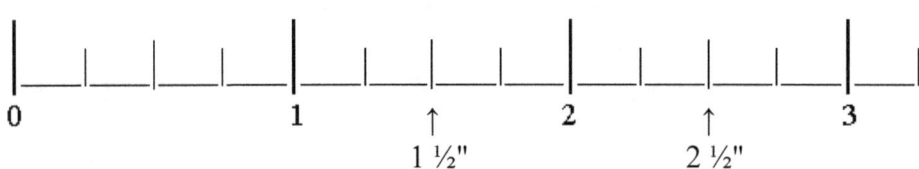

The tick marks on this ruler are at every eighth of an inch. The first tick mark is 1/8 of an inch. The second tick mark is 2/8 of an inch which is the same as 1/4 of an inch. The third tick mark is 3/8 of an inch. The next tick mark is 4/8 of an inch which is the same as 1/2 of an inch.

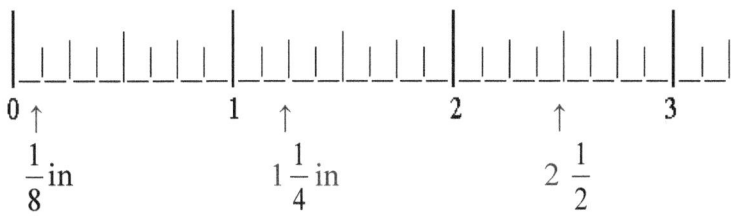

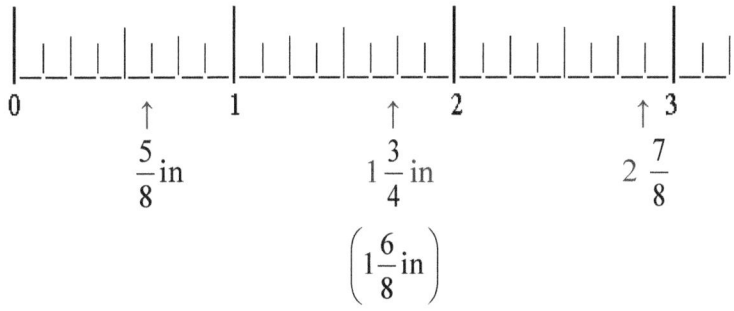

The tick marks on this ruler are at every sixteenth of an inch. The first tick mark is 1/16 of an inch. The second tick mark is 2/16 of an inch which is the same as 1/8 of an inch. The third tick mark is 3/16 of an inch. The next tick mark is 4/16 of an inch which is the same as 1/4 of an inch.

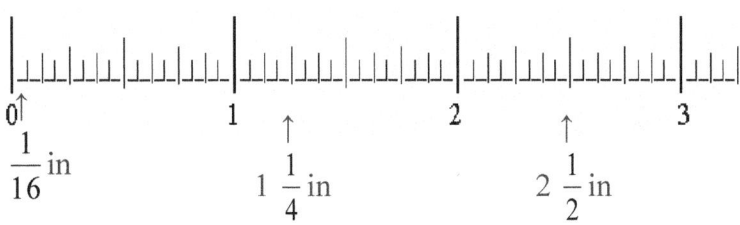

4.3 Homework: Reading a Ruler

Measure the following lines with the rulers given. If the end of the line does not fall exactly on a tick mark, then read the mark that is closest to the end of the line. Give answers as proper fractions or mixed numbers. Reduce fractions if possible.

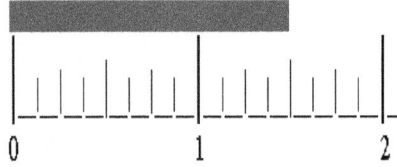

1. Answer: _____

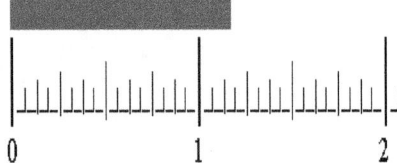

2. Answer: _____

3. Answer: _____

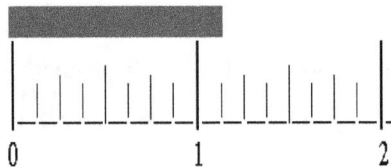

4. Answer: _____

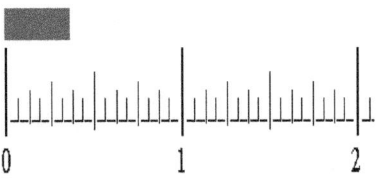

5. Answer: _____

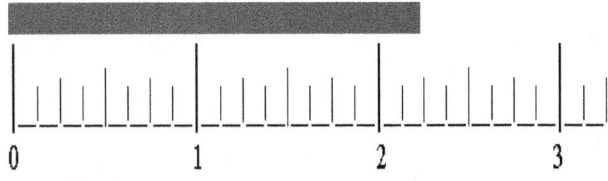

6. Answer: _____

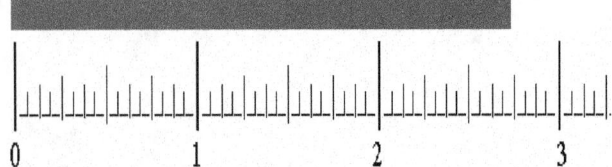

7. Answer: _____

Practice Test 4

Find the prime factorizations of the following numbers.

1. 135

2. 588

List all of the factors of the following numbers.

3. 45

4. 72

Find the LCM for the following groups of numbers.

5. 20 and 28

6. 14 and 26

7. Identify the proper fractions, improper fractions, and mixed numbers in the following group.

$$\left\{ \frac{8}{9}, \frac{7}{3}, 3\frac{1}{5}, \frac{9}{16}, \frac{18}{5}, \frac{15}{15}, 23\frac{5}{9} \right\}$$

Simplify the following if possible.

8. $\dfrac{253}{11}$

9. $\dfrac{17}{0}$

10. $\dfrac{444}{12}$

Practice Test 4 cont.

Simplify the following if possible.

11. $\dfrac{0}{13}$

12. $\dfrac{40}{56}$

13. $\dfrac{24}{120}$

14. $\dfrac{260}{390}$

15. $\dfrac{51}{84}$

16. $\dfrac{87}{145}$

17. Three out of fifteen residents have internet. What fraction of the residents have internet? Reduce if possible.

18. Last year, Hesperia High School has 450 graduates. If 250 of the graduates went to college this year, what fraction of the students went to college this year? Reduce if possible.

Convert the improper fractions to mixed numbers. (Reduce if possible.)

19. $\dfrac{17}{5}$

20. $\dfrac{58}{11}$

21. $\dfrac{74}{9}$

22. $\dfrac{30}{8}$

23. $\dfrac{85}{15}$

24. $\dfrac{100}{24}$

Convert the mixed numbers to improper fractions.

25. $3\dfrac{2}{9}$

26. $17\dfrac{2}{5}$

27. $50\dfrac{2}{3}$

Measure the following lines with the rulers given. If the end of the line does not fall exactly on a tick mark, then read the mark that is closest to the end of the line. Give answers as proper fractions or mixed numbers. Reduce fractions if possible.

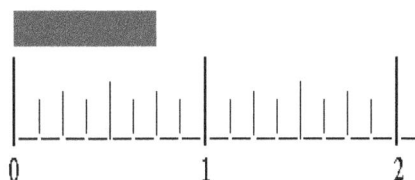

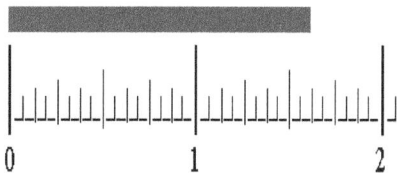

28. _____

29. _____

5.1 Multiplication with Fractions

We are now ready to learn how to perform operations with fractions. There are 4 basic operations; addition, subtraction, multiplication and division. Previously we learned how to perform these 4 operations with whole numbers. Now we will learn how to perform these operations with fractions.

To multiply two fractions

$$\frac{a}{b} \cdot \frac{c}{d} = \frac{a \cdot c}{b \cdot d}$$ In other words, multiply the numerators and multiply the denominators.

Example 1a. Multiply $\dfrac{3}{8} \cdot \dfrac{6}{7}$

Solution: $\dfrac{3}{8} \cdot \dfrac{6}{7} = \dfrac{18}{56}$ (Now we can reduce the fraction.)

The fraction can be reduced by either dividing both numerator and denominator by the GCF or by using the prime factorization method for reducing fractions. The prime factorization method will be shown in this example.

$$\frac{18}{56} = \frac{\cancel{2} \cdot 3 \cdot 3}{\cancel{2} \cdot 2 \cdot 2 \cdot 7} = \boxed{\frac{9}{28}}$$

Cross Reducing and Multiplying

Although the answer above is correct, it is <u>easier to reduce first before multiplying</u>. By reducing first, we will not have large fractions to reduce.

Procedure for Multiplying Fractions.

1. Reduce vertically or diagonally. Just don't reduce horizontally.

2. Then multiply across.

Example 1a again. Multiply: $\dfrac{3}{8} \cdot \dfrac{6}{7}$

Solution: Find a number that divides evenly into the numerator and the denominator.

2 will divide evenly into the 6 and the 8. Then multiply straight across.

$$\dfrac{3}{\underset{4}{\cancel{8}}} \cdot \dfrac{\overset{3}{\cancel{6}}}{7} = \boxed{\dfrac{9}{28}}$$

Example 1b. Simplify: $\dfrac{15}{16} \cdot \dfrac{12}{25}$

Solution: Find a number that divides evenly into the numerator and the denominator.

4 will divide evenly into the 12 and the 16. Also, 5 will divide evenly into the 15 and 25. Then multiply straight across.

$$\dfrac{\overset{3}{\cancel{15}}}{\underset{4}{\cancel{16}}} \cdot \dfrac{\overset{3}{\cancel{12}}}{\underset{5}{\cancel{25}}} = \boxed{\dfrac{9}{20}}$$

Prime Factorization Method for Multiplying Fractions

Many students prefer the prime factorization method because it is more procedural. You don't have to be thinking "What number divides evenly into these two numbers?"

Procedure: To Multiply Fractions using Prime Factorization

1. Rewrite the fraction using the prime factorizations.

2. Reduce the common factors. (that is, line-out one number on top with the same number on the bottom).

3. Multiply the remaining factors.

Example 1c. Simplify: $\dfrac{57}{77} \cdot \dfrac{22}{38}$

Solution: Find the prime factorization of each number. Write each in the same position the number is in each fraction. (Use tree method or division method for the prime factorization.) Keep in mind the divisibility rules.

```
   57            77            22            38
  /  \          /  \          /  \          /  \
 3   19        7   11        2   11        2   19
```

$\dfrac{3 \cdot 19}{7 \cdot 11} \cdot \dfrac{2 \cdot 11}{2 \cdot 19}$ Then line-out the common factors. The 19's cancel, the 11's cancel, and the 2's cancel. The 2's because you can reduce vertically.

$\dfrac{3 \cdot \cancel{19}}{7 \cdot \cancel{11}} \cdot \dfrac{\cancel{2} \cdot \cancel{11}}{\cancel{2} \cdot \cancel{19}}$ The only factor in the numerator is 3 (except for 1).
The only factor in the denominator is 7 (except for 1).

Answer: $\boxed{\dfrac{3}{7}}$

Your Turn Problem #1

Multiply the following:

a) $\dfrac{8}{11} \cdot \dfrac{7}{12}$

b) $\dfrac{51}{55} \cdot \dfrac{44}{85}$

Example 2. Simplify: $\dfrac{23}{45} \cdot \dfrac{15}{46}$

Solution: Find the prime factorization of each number. Write each in the same position the number is in each fraction. (Use tree method or division method for the prime factorization.) Keep in mind the divisibility rules.

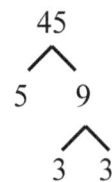

$$\frac{23}{3\cdot3\cdot5} \cdot \frac{3\cdot5}{2\cdot23}$$ Then line-out the common factors. The 23's cancel, the 3's cancel, and the 5's cancel.

$$\frac{\cancel{23}}{\cancel{3}\cdot3\cdot\cancel{5}} \cdot \frac{\cancel{3}\cdot\cancel{5}}{2\cdot\cancel{23}}$$ The only factor in the numerator is 1.
Multiply remaining factors in denominator.

Answer: $\boxed{\dfrac{1}{6}}$

Your Turn Problem #2

Multiply the following:

a) $\dfrac{18}{20} \cdot \dfrac{5}{27}$

b) $\dfrac{12}{75} \cdot \dfrac{25}{36}$

Example 3. Simplify: $\dfrac{56}{10} \cdot \dfrac{15}{21}$

Solution: Find the prime factorization of each number. Write each in the same position the number is in each fraction. (Use tree method or division method for the prime factorization.) Keep in mind the divisibility rules.

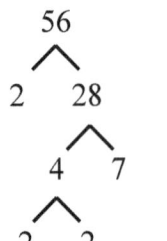

56
2   28
4   7
2   2

10
2   5

15
3   5

21
3   7

$\dfrac{2 \cdot 2 \cdot 2 \cdot 7}{2 \cdot 5} \cdot \dfrac{3 \cdot 5}{3 \cdot 7}$ Then line-out the common factors. 2's, 3's, 5's and 7's cancel.

$\dfrac{\cancel{2} \cdot 2 \cdot 2 \cdot \cancel{7}}{\cancel{2} \cdot \cancel{5}} \cdot \dfrac{\cancel{3} \cdot \cancel{5}}{\cancel{3} \cdot \cancel{7}}$ The only factor in the denominator is 1.
Multiply remaining factors in numerator.

$\dfrac{4}{1}$

Answer: $\boxed{4}$

Your Turn Problem #3

Multiply the following:

a) $\dfrac{24}{15} \cdot \dfrac{45}{18}$

b) $\dfrac{81}{15} \cdot \dfrac{70}{27}$

Multiplying a Fraction by a Whole Number

When multiplying a fraction and a whole number, write the whole number as a fraction by writing it with a denominator of 1.

Any number over 1 is itself. Examples: $\dfrac{5}{1} = 5$, $\dfrac{12}{1} = 12$

Example 4. Simplify: $24 \cdot \dfrac{5}{18}$ (do not leave an improper fraction for an answer.)

Solution: Write a 1 under the 24. Then reduce and multiply.

$$\frac{24}{1} \cdot \frac{5}{18} \qquad \text{Then reduce.}$$

$$\frac{\cancel{2} \cdot 2 \cdot 2 \cdot \cancel{3}}{1} \cdot \frac{5}{\cancel{2} \cdot \cancel{3} \cdot 3} = \frac{20}{3}$$

Since this is an improper fraction, change it to a mixed number.

Answer: $\boxed{6\dfrac{2}{3}}$

Your Turn Problem #4

Multiply the following. Do not leave improper fractions for answers.

a) $27 \cdot \dfrac{5}{36}$

b) $25 \cdot \dfrac{3}{5}$

Multiplying Mixed Numbers

Procedure for Multiplying with Mixed Numbers.
1. Convert all mixed numbers to improper fractions.
2. Reduce and multiply.
3. Then convert the answer back into a mixed number if it is improper.

Note: Depending on the directions, it may not always be necessary to change the improper fraction to a mixed number. However, if the answer was an answer to a word problem, it wouldn't sound right as an improper fraction. 4 ½ hours makes more sense than 9/2 hours.

Example 5. Simplify $5\frac{1}{2} \cdot 2\frac{1}{2}$ (Do not leave an improper fraction for an answer.)

Solution: Convert both mixed numbers to improper fractions.

$$\frac{11}{2} \cdot \frac{5}{2}$$ Since it doesn't reduce, just multiply straight across.

$$\frac{55}{4} = \boxed{13\frac{3}{4}}$$ This is an improper fraction. Change it to a mixed number

Your Turn Problem #5
Multiply the following. Do not leave improper fractions for answers.
a) $7\frac{1}{2} \cdot 2\frac{1}{2}$ b) $2\frac{2}{5} \cdot 3\frac{1}{3}$

Multiplying With More Than Two Fractions or Mixed Numbers

When multiplying with more than two fractions or mixed numbers.

1. If one or more of the fractions are mixed numbers, convert them to improper fractions.

2. Reduce <u>any</u> numerator with <u>any</u> denominator.

3. Multiply remaining factors.

4. Convert answer into a mixed number if possible.

Example 6. Simplify: $1\dfrac{22}{27} \cdot \dfrac{15}{16} \cdot \dfrac{72}{91}$ (Do not leave an improper fraction for an answer.)

Convert any mixed numbers to improper fractions.

Solution: $\dfrac{49}{27} \cdot \dfrac{15}{16} \cdot \dfrac{72}{91}$

Write fractions using each numbers prime factorization.

$$\dfrac{7 \cdot 7}{3 \cdot 3 \cdot 3} \cdot \dfrac{3 \cdot 5}{2 \cdot 2 \cdot 2 \cdot 2} \cdot \dfrac{2 \cdot 2 \cdot 2 \cdot 3 \cdot 3}{7 \cdot 13} \quad \text{(91 is not prime.)}$$

$$\dfrac{\cancel{7} \cdot 7}{\cancel{3} \cdot \cancel{3} \cdot \cancel{3}} \cdot \dfrac{\cancel{3} \cdot 5}{\cancel{2} \cdot \cancel{2} \cdot \cancel{2} \cdot 2} \cdot \dfrac{\cancel{2} \cdot \cancel{2} \cdot \cancel{2} \cdot \cancel{3} \cdot \cancel{3}}{\cancel{7} \cdot 13} = \dfrac{35}{26} = \boxed{1\dfrac{9}{26}}$$

Your Turn Problem #6

Multiply the following. Do not leave improper fractions for answers.

a) $3\dfrac{1}{6} \cdot \dfrac{1}{25} \cdot 1\dfrac{7}{38}$

b) $1\dfrac{1}{8} \cdot 4 \cdot \dfrac{5}{6}$

"Product"

To find the product of two numbers:
1. Rewrite the numbers in the same order as presented, in the sentence and place a multiplication sign in between.
2. Perform necessary steps of multiplication with fractions and mixed numbers.

Example 7. Find the product of $3\frac{1}{2}$ and 8.

Solution: 1. Rewrite: $3\frac{1}{2} \cdot 8$

2. Convert the $3\frac{1}{2}$ to an improper fraction and write a 1 under the 8.

$$\frac{7}{2} \cdot \frac{8}{1}$$

3. Then reduce and multiply.

$$\frac{7}{\cancel{2}_1} \cdot \frac{\cancel{8}^4}{1} = \frac{28}{1}$$

Answer: $\boxed{28}$

Your Turn Problem #7
Find the product of $\frac{5}{8}$ and $\frac{2}{5}$.

The word "of"

The word "of" indicates multiplication if a fraction precedes it.

Example 8. What is $\dfrac{3}{4}$ of 36?

Solution: Write a 1 under the 36. Then reduce and multiply.

$$\dfrac{3}{\overset{}{\underset{1}{4}}} \cdot \dfrac{\overset{9}{\cancel{36}}}{1} = \dfrac{27}{1} \boxed{27}$$

Answer: $\boxed{27}$

Your Turn Problem #8

What is $\dfrac{2}{9}$ of 720?

Word Problems involving fractions and multiplication.

General Steps to Solving Word Problems

1. Read and reread the problem carefully making note of all data (numbers), and keywords. Write down the information. Make a chart, picture, or diagram if possible. Write down the formula being used if applicable.

2. Identify the question. What is being asked?

3. Identify the operation to be used (addition, subtraction, multiplication, or division).

4. Do the math.

5. Answer the question in a sentence. Make sure proper units ($, feet, hours, books, etc.) are used and the answer makes sense.

Area of a Rectangle

The area of a rectangle can be found by multiplying the length and the width.

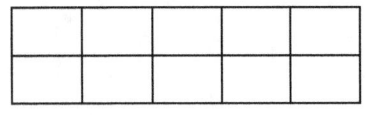

W = width
L = length

If the width of a rectangle is 2 feet and the length is 5 feet, then the area would be 10 square feet.

$A = L\ W$

$A = 2\,\text{ft} \times 5\ \text{ft}$

$A = 10\,\text{ft}^2$ (Area is usually written with square units: $\text{ft}^2, \text{mi}^2, \text{km}^2, \text{etc.}$)

Example 9. Find the area of a rectangle if the length is $2\frac{1}{2}$ ft and the width is $1\frac{3}{4}$ ft.

Solution: The formula for area of a rectangle: $A = L \cdot W$ (length times width)

Using the formula, multiply the length and width.

$A = 2\frac{1}{2}\,\text{ft} \times 1\frac{3}{4}\,\text{ft}$

$A = \frac{5}{2}\,\text{ft} \times \frac{7}{4}\,\text{ft}$

$A = \frac{35}{8}\,\text{ft}^2 = \boxed{4\frac{3}{8}\,\text{ft}^2}$

Your Turn Problem #9

Find the area of a rectangle if the length is $5\frac{1}{3}$ ft and the width is $3\frac{3}{4}$ ft.

Area of a Parallelogram

A parallelogram is a 4-sided shape formed by two pairs of parallel lines. Opposite sides are equal in length and opposite angles are equal in measure. To find the area of a parallelogram, multiply the base by the height. The formula is:

$$\boxed{\text{Area of a parallelogram} = \text{base} \times \text{height}}$$

Note: parallel lines are lines that are sketched in the exact same direction.

Example of a parallelogram

Example 10. Find the area of a parallelogram if the base is $7\frac{1}{2}$ ft and the height is $3\frac{1}{3}$ ft.

Solution: The formula for area of a parallelogram: $A = b \cdot h$ (base times height)

Using the formula, multiply the base and height.

$$A = 7\frac{1}{2}\text{ft} \times 3\frac{1}{3}\text{ft}$$

$$A = \frac{15}{2}\text{ft} \times \frac{10}{3}\text{ft} \quad \text{(reduce and multiply)}$$

$$A = \boxed{25\,\text{ft}^2}$$

Your Turn Problem #10

Find the area of a parallelogram if the base is $4\frac{1}{3}$ ft and the height is $1\frac{1}{2}$ ft.

Area of a Triangle

The area of a triangle is given by the formula: $A = \dfrac{1}{2} \cdot b \cdot h$, where b = base and h = height.

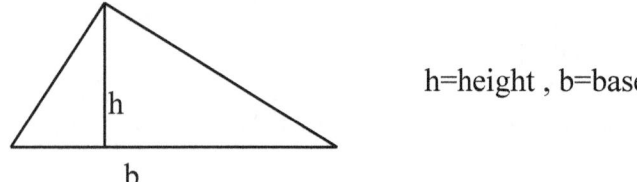

h=height , b=base

Note: There is a reason why this formula for area is (base × height) ÷ 2. Any triangle is half of a parallelogram. Therefore, if we only want half of the parallelogram, we divide the area of a parallelogram by 2 to find the area of a triangle.

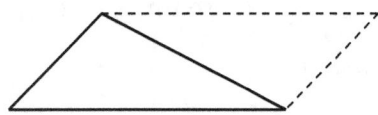

Any triangle is one-half of a parallelogram.

Example 11. Find the area of the triangle if the base is 20 feet and the height is $2\frac{3}{4}$ ft.

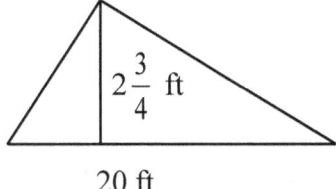

20 ft

Solution: The formula for area of a triangle: $A = \frac{1}{2} \cdot b \cdot h \ \left(\frac{1}{2} \times base \times height \right)$.

$$A = \frac{1}{2} \times 20\text{ft} \times 2\frac{3}{4}\text{ft}$$

$$A = \frac{1}{2}\text{ft} \times \frac{20}{1}\text{ft} \times \frac{11}{4}\text{ft} \ \text{(Reduce and multiply.)}$$

$$A = \boxed{27\frac{1}{2}\text{ft}^2}$$

Your Turn Problem #11

Find the area of the triangle if the base is 15 meter and the height is $4\frac{1}{2}$ meters.

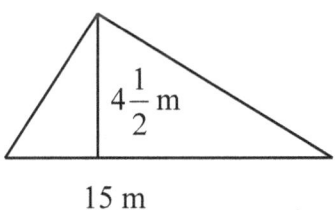

15 m

Answer:_____

227

Example 12. Three-fifths of the graduating students at Victor Valley College will transfer to a university. If there are 935 graduating students, how many will transfer to a university?

Solution: Three-fifths can be written as a fraction: $\dfrac{3}{5}$. After the fraction is the word "of". "Of" usually implies multiplication if a fraction precedes it. The number of graduating students is 935. So we have $\dfrac{3}{5} \cdot 935$.

$$\dfrac{3}{\overset{}{\underset{1}{\cancel{5}}}} \cdot \dfrac{\overset{187}{\cancel{935}}}{1} = \dfrac{561}{1} = 561$$

Answer: | VVC will have 561 graduating students that transfer to a university. |

Note: The numerator of the second fraction, 935, is a large number. It is not practical to find the prime factorization for 935. It is easier to cross reduce in this case.

Also note: If we multiplied straight across before reducing, this would work out fine as well.

$$\dfrac{3}{5} \cdot \dfrac{935}{1} = \dfrac{2805}{5} \quad \text{(now calculate } 2805 \div 5\text{)}$$
$$= 561$$

Your Turn Problem #12

Four-fifteenths of Lori's monthly check goes to rent. If her monthly check is $3000, how much is her rent?

5.1 Homework: Multiplication with Fractions

Find each product in lowest terms. Do not leave improper fractions for answers.

1. $\dfrac{3}{5} \cdot \dfrac{9}{15}$

2. $\dfrac{12}{25} \cdot \dfrac{15}{16}$

3. $\dfrac{11}{15} \cdot \dfrac{3}{22}$

4. $\dfrac{3}{5} \cdot \dfrac{5}{10}$

5. $\dfrac{2}{9} \cdot 12$

6. $\dfrac{3}{5} \cdot \dfrac{12}{14}$

7. $\dfrac{26}{9} \cdot \dfrac{2}{13}$

8. $1\dfrac{2}{3} \cdot \dfrac{3}{5}$

5.1 Homework: Multiplication with Fractions cont.

Find each product in lowest terms. Do not leave improper fractions for answers.

9. $3\dfrac{1}{5} \cdot 1\dfrac{1}{24}$

10. $2\dfrac{1}{3} \cdot 3\dfrac{1}{2}$

11. $5\dfrac{1}{3} \cdot \dfrac{7}{8}$

12. $3\dfrac{4}{9} \cdot 12$

13. $6\dfrac{2}{5} \cdot \dfrac{15}{19}$

14. $5\dfrac{2}{7} \cdot 3\dfrac{1}{7}$

15. $\dfrac{7}{12} \cdot \dfrac{9}{16} \cdot \dfrac{5}{14}$

16. $\dfrac{18}{25} \cdot \dfrac{3}{22} \cdot \dfrac{55}{63}$

5.1 Homework: Multiplication with Fractions cont.

Find each product in lowest terms. Do not leave improper fractions for answers.

17. $\dfrac{5}{18} \cdot \dfrac{3}{4} \cdot 12$

18. $\dfrac{21}{25} \cdot \dfrac{10}{12} \cdot \dfrac{15}{28}$

19. $\dfrac{28}{33} \cdot \dfrac{15}{91} \cdot \dfrac{22}{30}$

20. $\dfrac{23}{25} \cdot 8 \cdot \dfrac{45}{46}$

21. $\dfrac{29}{58} \cdot 12 \cdot \dfrac{5}{30}$

22. $\dfrac{7}{90} \cdot \dfrac{5}{84}$

23. $\dfrac{7}{24} \cdot \dfrac{5}{20} \cdot 8$

24. $\dfrac{27}{35} \cdot \dfrac{7}{36} \cdot \dfrac{25}{11}$

5.1 Homework: Multiplication with Fractions cont.

Find each product in lowest terms. Do not leave improper fractions for answers.

25. $\dfrac{24}{100} \cdot \dfrac{36}{48} \cdot 1\dfrac{6}{9}$

26. $9\dfrac{3}{8} \cdot \dfrac{16}{36} \cdot 2\dfrac{13}{25}$

27. $\dfrac{57}{68} \cdot \dfrac{51}{115} \cdot \dfrac{69}{76}$

28. $4\dfrac{2}{3} \cdot 3\dfrac{1}{2} \cdot 5\dfrac{1}{4}$

29. $4\dfrac{1}{4} \cdot 12 \cdot 3\dfrac{1}{2}$

30. Find $\dfrac{2}{3}$ of 3672.

31. Find $\dfrac{7}{15}$ of 5175.

32. What is $\dfrac{1}{3}$ of $3\dfrac{4}{9}$?

5.1 Homework: Multiplication with Fractions cont.

Solve each applications. Do not leave improper fractions for answers.

33. Find the product of $2\frac{1}{3}$ and $3\frac{1}{2}$.

34. Find the product of $\frac{5}{16}$ and 8064.

35. Find the height of a stack of 24 shoe boxes if each shoe box is seven-eighths of a foot high.

36. $\frac{3}{4}$ of a co-ed soccer team are women. If there are 32 players on the team, how many are women?

37. Four-fifteenths of Sammy's monthly paycheck goes to his mortgage payment. If his monthly paycheck is $2,775, how much is his mortgage payment?

5.1 Homework: Multiplication with Fractions cont.

Solve each applications. Do not leave improper fractions for answers.

38. A walkway needs $18\frac{1}{2}$ ft^2 of brick to cover it completely. If the owner only wants to cover

$\frac{3}{4}$ of the walkway, how much brick does he need?

39. A kitchen counter has dimensions of $3\frac{1}{4}$ ft by $8\frac{5}{12}$ ft. What is the area of counter top?

40. A race is $4\frac{1}{2}$ miles long. If Tim has completed $\frac{1}{2}$ of the race, how many miles has he run?

41. A recipe for French Toast requires $\frac{3}{4}$ cup of batter for each serving. If 5 servings are to be

served for breakfast, how much batter is needed?

5.1 Homework: Multiplication with Fractions cont.

Solve each applications. Do not leave improper fractions for answers.

42. A driveway requires $4\frac{5}{6}$ cubic yards of concrete. If the contractor is asked to enlarge the

driveway to $2\frac{1}{2}$ times its current size, how much concrete will she need?

43. Find the area of a parallelogram if the base is $6\frac{2}{3}$ ft and the height is $1\frac{1}{5}$ ft ?

44. Find the area of a triangle if the base is 18 ft and the height is $4\frac{2}{3}$ ft ?

45. Find the area of a triangle if the base is $8\frac{1}{2}$ ft and the height is $3\frac{1}{2}$ ft ?

5.2 Division with Fractions

Reciprocal

The reciprocal of a number is: what must be multiplied by the number to get a product of "1".

Example: The reciprocal of $\dfrac{3}{4}$ is $\dfrac{4}{3}$ because $\dfrac{3}{4} \cdot \dfrac{4}{3} = 1$

So the reciprocal is the fraction inverted. The numerator and the denominator are interchanged.

Examples:

Fraction	Reciprocal	
$\dfrac{7}{11}$	$\dfrac{11}{7}$	
$\dfrac{1}{9}$	$\dfrac{9}{1}$	
12	$\dfrac{1}{12}$	$\left(12 = \dfrac{12}{1}\right)$

The only number that does not have a reciprocal is 0. The reciprocal of $\dfrac{0}{1}$ is $\dfrac{1}{0}$ which is undefined.

Division of Fractions

Procedure: Dividing a fraction by another fraction.

Step 1. Change any mixed numbers to improper fractions.

Step 2. Rewrite the problem changing the division sign to a multiplication sign and inverting any fraction that originally followed a division sign into its reciprocal. The first fraction stays the same, the second fraction is inverted.

Step 3. Follow the procedures of multiplying fractions--reduce if possible and multiply.

Step 4. If improper, convert it to a mixed number.

Example 1: Simplify $12 \div \dfrac{1}{2}$

Solution: Write a 1 under the whole number and change to multiplication.

$$\frac{12}{1} \div \frac{1}{2} \quad \Rightarrow \quad \frac{12}{1} \times \frac{2}{1} = \frac{24}{1} = \boxed{24}$$

Does the answer seem correct? Division is the process that asks, "How many of something are contained in another?" For example, $56 \div 8$, asks "how many 8's are contained in 56?" In our problem $12 \div \dfrac{1}{2}$ asks, "how many 1/2's are contained in 12?" How many half-dollars are in 12 dollars? There are 24 half-dollars in 12 dollars.

Your Turn Problem #1

Simplify: $\dfrac{12}{25} \div \dfrac{9}{10}$

Example 2: Simplify $5\dfrac{1}{10} \div 2\dfrac{4}{15}$. Do not leave an improper fraction for an answer.

Solution: Convert mixed numbers to improper. Change to multiplication. Reduce and multiply.

$$5\frac{1}{10} \div 2\frac{4}{15} \quad \Rightarrow \quad \frac{51}{10} \div \frac{34}{15}$$

$$\frac{51}{10} \cdot \frac{15}{34}$$

$$\frac{3 \cdot \cancel{17}}{2 \cdot \cancel{5}} \cdot \frac{3 \cdot \cancel{5}}{2 \cdot \cancel{17}} = \frac{9}{4} = \boxed{2\frac{1}{4}}$$

Your Turn Problem #2

Simplify: $1\dfrac{11}{15} \div 3\dfrac{9}{10}$

Division is an operation where an amount is being divided into groups. For example, if you have $20 to divide among 4 people, how much would each person get?

  Answer: $20 \div 4 = \$5$

Note: **The total must be written first.**

Division is not commutative. $20 \div 4$ is not the same as $4 \div 20$.

Example 3. A certain size bottle holds exactly 2/3 pints of liquid. How many of these bottles can be filled from a 12-pint container?

Solution: This is a division problem because a quantity is being separated into groups. Remember to write the total quantity first.

$$12 \div \dfrac{2}{3}$$

Change to multiplication, reduce and multiply.

$$\dfrac{\overset{6}{\cancel{12}}}{1} \cdot \dfrac{3}{\underset{1}{\cancel{2}}} = 18$$

Answer: $\boxed{\text{18 bottles can be filled.}}$

Your Turn Problem #3

A 20 ft pipe must be cut into pieces $1\frac{3}{4}$ ft long. How many $1\frac{3}{4}$ ft pieces of pipe can be

obtained? (Hint: The answer is how many *full pieces* you will have that are $1\frac{3}{4}$ ft long.)

Example 4. A piece of land measures $6\frac{3}{4}$ acres and is for a sale at a price $216,000. What is

the price per acre?

Solution: Key Word – "per". The word "per" usually indicates division.

Remember to write the total quantity first.

$$216,000 \div 6\frac{3}{4}$$

Convert mixed number to an improper fraction

$$\frac{216,000}{1} \div \frac{27}{4}$$

Now change to multiplication, then simplify.

$$\frac{216,000}{1} \cdot \frac{4}{27} = \frac{864,000}{27} = 32,000$$

Answer: The price is $32,000 per acre.

Your Turn Problem #4

A piece of land measures $2\frac{2}{3}$ acres and is for sale at a price of $168,000. What is the price per

acre?

5.2 Homework: Division with Fractions

Divide the following. Do not leave improper fractions for answers.

1. $\dfrac{3}{5} \div \dfrac{9}{15}$

2. $\dfrac{12}{25} \div \dfrac{16}{15}$

3. $\dfrac{11}{15} \div \dfrac{2}{5}$

4. $\dfrac{4}{9} \div \dfrac{3}{4}$

5. $\dfrac{9}{13} \div \dfrac{9}{26}$

6. $\dfrac{3}{4} \div 12$

7. $\dfrac{6}{13} \div \dfrac{5}{16}$

8. $\dfrac{7}{12} \div \dfrac{9}{10}$

9. $\dfrac{18}{25} \div \dfrac{9}{10}$

10. $18 \div \dfrac{2}{3}$

11. $\dfrac{7}{12} \div \dfrac{3}{8}$

12. $\dfrac{21}{16} \div \dfrac{28}{5}$

5.2 Homework: Division with Fractions cont.

Divide the following. Do not leave improper fractions for answers.

13. $\dfrac{12}{77} \div \dfrac{18}{33}$

14. $\dfrac{49}{60} \div \dfrac{21}{50}$

15. $\dfrac{23}{40} \div \dfrac{46}{8}$

16. $\dfrac{29}{17} \div \dfrac{58}{34}$

17. $1\dfrac{2}{3} \div 1\dfrac{1}{2}$

18. $5\dfrac{1}{3} \div 2\dfrac{2}{5}$

19. $2\dfrac{1}{3} \div 3\dfrac{1}{2}$

20. $4\dfrac{2}{3} \div 2\dfrac{1}{3}$

5.2 Homework: Division with Fractions cont.

Divide the following. Do not leave improper fractions for answers.

21. $3\frac{3}{4} \div \frac{5}{9}$

22. $2\frac{1}{2} \div 4\frac{1}{2}$

23. $3\frac{1}{2} \div 2\frac{4}{5}$

24. $4\frac{2}{3} \div 12$

Solve each application. Do not leave improper fractions for answers.

25. A manufacturer has 45 ½ yards of material. If a shirt requires 1 ¾ yards, how many shirts can be made?

26. A certain size bottle holds exactly 4/5 pint of liquid. How many of these bottles can be filled from a 20-pint container?

5.2 Homework: Division with Fractions cont.

Solve each application. Do not leave improper fractions for answers.

27. Henry has 41 ½ ft of pipe. He wants to cut it into pieces 3 ¾ ft long. How many pieces that are 3 ¾ ft long of pipe will he have?

28. How many $4\frac{1}{2}$ -in pieces of pipe, can be cut from a 75-in long piece of pipe?

29. A piece of land measures $3\frac{4}{5}$ acres and is for sale at a price of \$129,200. What is the price per acre?

30. A liquid sample which weighs $5\frac{3}{4}$ ml (milliliters) is to be separated into 2 different vials.

How much of the liquid will be in each vial?

5.3 Solving Equations of the Form a · x = b

Variable: Often, we use letters to represent an unknown number. Letters like x and y are most commonly used. This letter is called a variable.

Variable Term: A variable term is a product of a number and a variable(s).

Examples of variable terms: $7 \cdot x$, $\dfrac{2}{3} \cdot x$, $6 \cdot y$

When writing the product of a number and a variable, it is common to not write the multiplication symbol. So when we see a number and a variable next to each other, the operation is assumed to be multiplication.

$$7 \cdot x, \quad \frac{2}{3} \cdot x, \quad 6 \cdot y \;\Rightarrow\; 7x, \quad \frac{2}{3}x, \quad 6y$$

Constant Term: A constant term is another description of a number without the variable. A constant term is also called a numerical expression.

Examples of constant terms: 12, $\dfrac{8}{7}$, $2\dfrac{1}{2}$

Expressions: An expression is a term or a sum or difference of terms that may use numbers, variables, or both.

Examples of expressions: $3x$, $2x+5$, $8x-7$, $6y$

Equation: An equation is used to express the equality of two variable or numerical expressions. An equation is a statement where two equal expressions are separated by an " = " (equal) sign.

Examples of equations: $5 \cdot x = 40$, $\dfrac{2}{3} \cdot x = \dfrac{9}{11}$, $3\dfrac{1}{2}x = 2\dfrac{2}{3}$

The first example above, $5 \cdot x = 40$, is asking: "Five times what number is 40?" The answer is 8. Sometimes we can come up with the answer in our head. This is not always true. The second example of an equation, $\dfrac{2}{3} \cdot x = \dfrac{9}{11}$, is asking "Two-thirds times what number is nine-elevenths?"

Most of us cannot come up with the answer in our head. We then need rules and procedures to follow so that we can solve this equation.

Basic Equation: An equation of the form: variable = constant

   Examples: $x = 5$, $y = 7$, $a = 0$

Multiplication and Division Properties of Equality

If two expressions are equal to each other, then you can multiply or divide by the exact same non-zero constant to both sides. The two sides will remain equal.

Solving Equations of the form a · x = b

To solve the equation, $a \cdot x = b$, where a and b are numbers and x is an unknown.

1. Our goal is to get x by itself. We want to obtain the basic equation of x = #.

2. To get x by itself we can divide by 'a' on both sides to get x by itself.

 Note: We just use the letter x to represent an unknown value. The letter could be any other letter of the alphabet.

Example 1a. Solve: $5 \cdot x = 40$

Solution: Our goal is to get the x by itself. To get x by itself, divide by 5 on both sides.

    Draw a line under each side to indicate division.

    Write the number that is in front of the x underneath each line.

$$\frac{5 \cdot x}{5} = \frac{40}{5}$$

Answer: $\boxed{x = 8}$

Example 1b. Solve: $9 \cdot x = 423$

Solution: Our goal is to get the x by itself. To get x by itself, divide by 9 on both sides.

    Draw a line under each side to indicate division.

    Write the number that is in front of the x underneath each line.

$$\frac{9 \cdot x}{9} = \frac{423}{9}$$

Answer: $\boxed{x = 47}$

Your Turn Problem #1

Solve: $12 \cdot x = 540$

Answer:_____

Answers are not always whole numbers. They can be fractions, mixed numbers, or decimals.

Example 2. Solve: $8 \cdot x = 255$

Solution: Our goal is to get the x by itself. To get x by itself, divide by 8 on both sides.

Draw a line under each side to indicate division.

Write the number that is in front of the x underneath each line.

$$\frac{8 \cdot x}{8} = \frac{255}{8}$$

Answer: $\boxed{x = 31\frac{7}{8}}$

$$\begin{array}{r} 31 \\ 8\overline{)255} \\ -24 \\ \hline 15 \\ -8 \\ \hline 7 \end{array}$$

Your Turn Problem #2

Solve: $11 \cdot x = 90$

Answer:_____

Example 3. Solve: $30 \cdot x = 28$

Solution: Our goal is to get the x by itself.

To get x by itself, divide by 30 on both sides. Draw a line under each side to indicate division. Write the number that is in front of the x underneath each line.

$\dfrac{30 \cdot x}{30} = \dfrac{28}{30}$ (Since 30 does not "go into" 28, reduce the fraction.)

Answer: $\boxed{x = \dfrac{14}{15}}$ (Divide numerator and denominator by 2.)

Your Turn Problem #3

Solve: $40 \cdot x = 25$

Answer:_____

Note: A number multiplied by its reciprocal is equal to 1.

Examples: $\dfrac{2}{3} \cdot \dfrac{3}{2} = 1$, $\dfrac{5}{8} \cdot \dfrac{8}{5} = 1$

Solving Equations of the form $a \cdot x = b$ where a or b is a fraction.

To solve an equation where a or b is a fraction:

1. Multiply both sides of the equation by the reciprocal of 'a'.

2. Simplify the left hand side and the right hand side.

Example 4. Solve: $\dfrac{2}{3} \cdot x = 12$

Solution: Our goal is to get the x by itself. To get x by itself, multiply by $\dfrac{3}{2}$ on both sides.

$$\left(\dfrac{3}{2}\right) \cdot \dfrac{2}{3} \cdot x = 12 \cdot \left(\dfrac{3}{2}\right)$$

On the left hand side, the $\dfrac{3}{2}$ and the $\dfrac{2}{3}$ multiply to equal 1. So we just have an x on the left hand side.

On the right side we need to multiply the 12 and the $\dfrac{3}{2}$. Write a 1 under the 12, reduce and multiply.

$$\left(\dfrac{\cancel{3}}{\cancel{2}}\right) \cdot \dfrac{\cancel{2}}{\cancel{3}} \cdot x = \dfrac{\overset{6}{\cancel{12}}}{1} \cdot \left(\dfrac{3}{\underset{1}{\cancel{2}}}\right)$$

Answer: $\boxed{x = 18}$

Your Turn Problem #4

Solve: $\dfrac{5}{8} \cdot a = 40$

Answer:_____

Example 5. Solve: $\dfrac{8}{15} \cdot x = \dfrac{11}{25}$

Solution: Our goal is to get the x by itself. To get x by itself, multiply by $\dfrac{15}{8}$ on both sides.

$$\left(\dfrac{15}{8}\right) \cdot \dfrac{8}{15} \cdot x = \dfrac{11}{25} \cdot \left(\dfrac{15}{8}\right)$$

On the left hand side, the $\dfrac{15}{8}$ and the $\dfrac{8}{15}$ multiply to equal 1. So we just have an x

on the left hand side.

On the right side, we need to multiply the $\dfrac{11}{25}$ and the $\dfrac{15}{8}$. Reduce and multiply.

$$\left(\dfrac{\cancel{15}}{\cancel{8}}\right) \cdot \dfrac{\cancel{8}}{\cancel{15}} \cdot x = \dfrac{11}{\underset{5}{\cancel{25}}} \cdot \left(\dfrac{\overset{3}{\cancel{15}}}{8}\right)$$

Answer: $\boxed{x = \dfrac{33}{40}}$

Your Turn Problem #5

Solve: $\dfrac{9}{16} \cdot y = \dfrac{21}{40}$

Answer:_____

Example 6. Solve: $8 \cdot x = \dfrac{4}{9}$

Solution: Whenever there is a fraction on either side, multiply by the reciprocal of the number in front of the variable on each side. In this case, multiply by $\dfrac{1}{8}$ on both sides.

$$\left(\frac{1}{8}\right) \cdot \frac{8}{1} \cdot x = \frac{4}{9} \cdot \left(\frac{1}{8}\right)$$

On the left hand side, the $\dfrac{8}{1}$ and the $\dfrac{1}{8}$ multiply to equal 1. So we just have an x on the left hand side.

On the right side we need to multiply the $\dfrac{4}{9}$ and the $\dfrac{1}{8}$. Reduce and multiply.

$$\left(\frac{1}{\cancel{8}}\right) \cdot \frac{\cancel{8}}{1} \cdot x = \frac{\overset{1}{\cancel{4}}}{9} \cdot \left(\frac{1}{\underset{2}{\cancel{8}}}\right)$$

Answer: $\boxed{x = \dfrac{1}{18}}$

Your Turn Problem #6

Solve: $12 \cdot x = \dfrac{8}{11}$

Answer:_____

Example 7. Solve: $3\frac{1}{2} \cdot x = 4\frac{2}{3}$

Solution: Whenever we are multiplying or dividing and there are mixed numbers, the first step is to change all mixed numbers to improper fractions. Then multiply by the reciprocal of the number in front of x on both sides.

$$\left(\frac{2}{7}\right) \cdot \frac{7}{2} \cdot x = \frac{14}{3} \cdot \left(\frac{2}{7}\right)$$

On the left hand side, the $\frac{2}{7}$ and the $\frac{7}{2}$ multiply to equal 1. So we just have an x on the left hand side.

On the right side we need to multiply the $\frac{14}{3}$ and the $\frac{2}{7}$. Reduce and multiply.

$$\left(\frac{\cancel{2}}{\cancel{7}}\right) \cdot \frac{\cancel{7}}{\cancel{2}} \cdot x = \frac{\overset{2}{\cancel{14}}}{3} \cdot \left(\frac{2}{\cancel{7}}\right)$$
$$\underset{1}{}$$

$$x = \frac{4}{3}$$

Answer: $\boxed{x = 1\frac{1}{3}}$

Your Turn Problem #7

Solve: $4\frac{2}{3} \cdot x = 5\frac{1}{2}$

Answer:_____

5.3 Homework: Solving Equations of the Form a · x = b
Solve each of the following equations. Do not leave improper fractions for answers.

1. $3 \cdot x = 231$

2. $9 \cdot a = 531$

3. $12 \cdot x = 2496$

4. $5 \cdot c = 735$

5. $8 \cdot h = 31$

6. $20 \cdot x = 88$

7. $15 \cdot x = 123$

8. $10 \cdot x = 25$

9. $55 \cdot x = 22$

10. $91 \cdot x = 49$

5.3 Homework: Solving Equations of the Form a · x = b cont.

Solve each of the following equations. Do not leave improper fractions for answers.

11. $51 \cdot d = 17$

12. $28 \cdot x = 7$

13. $\dfrac{2}{3} \cdot y = 12$

14. $\dfrac{3}{5} \cdot n = 27$

15. $\dfrac{5}{8} \cdot x = 11$

16. $\dfrac{2}{9} \cdot x = 7$

17. $\dfrac{3}{8} \cdot x = \dfrac{9}{10}$

18. $\dfrac{25}{36} \cdot x = \dfrac{15}{16}$

5.3 Homework: Solving Equations of the Form a · x = b cont.

Solve each of the following equations. Do not leave improper fractions for answers.

19. $\dfrac{11}{30} \cdot m = \dfrac{33}{40}$

20. $9 \cdot x = \dfrac{3}{5}$

21. $8 \cdot x = \dfrac{6}{11}$

22. $3\dfrac{2}{3} \cdot x = 55$

23. $2\dfrac{2}{5} \cdot g = 5\dfrac{1}{3}$

24. $2\dfrac{4}{5} \cdot x = 3\dfrac{1}{2}$

Practice Test 5

Perform the indicated operation. Do not leave improper fractions for answers.

1. $\dfrac{25}{36} \cdot \dfrac{24}{35}$

2. $\dfrac{7}{24} \cdot 18$

3. $15 \cdot \dfrac{3}{5}$

4. $3\dfrac{1}{3} \cdot \dfrac{3}{10}$

5. $7\dfrac{1}{3} \cdot \dfrac{9}{16}$

6. $2\dfrac{4}{5} \cdot 1\dfrac{7}{28}$

7. $\dfrac{25}{28} \div \dfrac{15}{16}$

8. $\dfrac{19}{26} \div \dfrac{38}{39}$

Practice Test 5 cont.

Perform the indicated operation. Do not leave improper fractions for answers.

9. $\dfrac{7}{12} \div 14$

10. $7\dfrac{1}{2} \div 3\dfrac{1}{3}$

11. $4\dfrac{1}{5} \div \dfrac{14}{15}$

12. $18 \div 2\dfrac{1}{6}$

Solve the following applications. Do not leave improper fractions for answers.

13. It is estimated that three-fifths of the graduates at VVC will attended a university the following year. If there are 725 graduates at VVC this year, how many of them will be going to a university next year?

14. Jennifer spends $\dfrac{2}{15}$ of her monthly income on her car payment. If her monthly income is $3,150, how much is her car payment?

Practice Test 5 cont.

Perform the indicated operation. Do not leave improper fractions for answers.

15. A race is $4\frac{1}{2}$ miles long. If Ray has completed $\frac{3}{4}$ of the race, how many miles has he run?

16. $\frac{3}{4}$ of a cup of sugar is needed for a cake. How much sugar is needed if the cook only wants to make half of a cake?

17. Find the area of a rectangle if the width is $2\frac{1}{3}$ ft and the length is $8\frac{1}{2}$ ft.

18. A certain size bottle holds exactly $\frac{3}{4}$ pint of liquid. How many of these bottles can be filled from a 60-pint container?

19. A manufacturer has $105\frac{1}{2}$ yards of material. If a shirt requires $1\frac{3}{4}$ yards, how many shirts can be made?

Practice Test 5 cont.

Perform the indicated operation. Do not leave improper fractions for answers.

20. Mike has 80 ft of cable. He wants to cut it into pieces $2\frac{1}{4}$ ft long. How many pieces that

are $2\frac{1}{4}$ ft long of cable will he have?

Solve the following Equations. Do not leave improper fractions for answers.

21. $12 \cdot x = 30$

22. $48 \cdot x = 16$

23. $\frac{4}{9} \cdot n = 36$

24. $\frac{18}{35} \cdot x = \frac{27}{50}$

25. $2\frac{2}{5} \cdot x = 18$

26. $6\frac{1}{2} \cdot x = 2\frac{3}{5}$

6.1 Reviewing the LCM and Equivalent Fractions

Reviewing the LCM (Least Common Multiple)

Please review section P.3 for procedures and more examples.

Example 1: Find the LCM of 15 and 18.

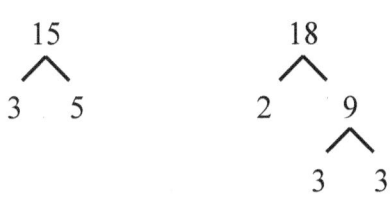

$$15 = 3 \cdot 5$$
$$18 = 2 \cdot 3 \cdot 3$$
$$\text{LCM} = 2 \cdot 3 \cdot 3 \cdot 5$$
$$= \boxed{90}$$

Step 1. Find the prime factorization of each number.

Step 2. Write LCM= below the prime factorization of each number.

Consider each prime number. Write it the greatest number of times it occurs in any one factorization.

2: it occurs once.

3: it occurs twice.

5: it occurs once.

Now multiply. This is the LCM.

Your Turn Problem #1

Find the LCM of 14 and 20

$\qquad$ 14 =

$\qquad$ 20 =

$\qquad$ LCM =

Equivalent Fractions

Recall the Multiplicative Identity (also called the multiplication property of 1): $a \cdot 1 = a$.

When we multiply a number by 1, we get the same number.

Example: $5 \cdot 1 = 5$

Recall, any fraction with the same numerator and denominator is equal to 1.

Examples: $\dfrac{3}{3} = 1$, $\dfrac{5}{5} = 1$, $\dfrac{5{,}278}{5{,}278} = 1$

If we take a fraction, such as $\dfrac{1}{2}$, and multiply the numerator and denominator by the same

number, we get an equivalent fraction.

$$\frac{1}{2} \cdot \frac{3}{3} = \frac{3}{6} \qquad\qquad \frac{1}{2} \cdot \frac{5}{5} = \frac{5}{10}$$

So $\dfrac{1}{2}, \dfrac{3}{6}$, and $\dfrac{5}{10}$ are all equivalent fractions. $\dfrac{3}{6}$, and $\dfrac{5}{10}$ both simplify to $\dfrac{1}{2}$.

As long as we multiply the numerator and denominator by the same number, we will always

obtain an equivalent fraction.

Example 2. Write the following fraction as an equivalent fraction with the given denominator.

$$\frac{5}{9} = \frac{\quad}{27}$$

Solution: We want an equivalent fraction with a denominator of 27. What number multiplied

by 9 will give 27 (or divide 27 by 3)? The answer is 3. Therefore, multiply the

denominator by 3 to get 27 and multiply the numerator by 3 to obtain the desired

equivalent fraction.

$$\frac{5}{9} \cdot \frac{3}{3} = \boxed{\frac{15}{27}}$$

Your Turn Problem #2

Write the following fraction as an equivalent fraction with the given denominator:

$$\frac{7}{12} = \frac{}{48}$$

Example 3. Write the following fraction as an equivalent fraction with the given denominator.

$$\frac{9}{14} = \frac{}{392}$$

Solution: We want an equivalent fraction with a denominator of 392. What number multiplied by 14 will give 392? In this case, it is easier to find the number by dividing 14 into 392.

$$\begin{array}{r} 28 \\ 14\overline{)392} \\ -28 \\ \hline 112 \\ -112 \\ \hline 0 \end{array}$$

Now multiply numerator and denominator by 28 to obtain the equivalent fraction with a denominator of 392.

$$\frac{9}{14} \cdot \frac{28}{28} = \boxed{\frac{252}{392}}$$

Your Turn Problem #3

Write the following fraction as an equivalent fraction with the given denominator:

$$\frac{3}{19} = \frac{}{969}$$

Example 4. Write the following fraction as an equivalent fraction with the given denominator.

$$8 = \frac{}{12}$$

Solution: First, write a "1" under the 8.

$$\frac{8}{1} = \frac{}{12}$$

We want an equivalent fraction with a denominator of 12. What number multiplied by 1 will give 12? Answer: 12.

Now multiply numerator and denominator by 12 to obtain the equivalent fraction with a denominator of 12.

$$\frac{8}{1} \cdot \frac{12}{12} = \boxed{\frac{96}{12}}$$

Your Turn Problem #4

Write the following fraction as an equivalent fraction with the given denominator:

$$14 = \frac{}{5}$$

Example 5. Write the following fraction as an equivalent fraction with the given denominator.

$$\frac{3}{16} = \frac{}{16}$$

Solution: We want an equivalent fraction with a denominator of 16. What number multiplied by 16 will give 16? Answer: 1.

Now multiply numerator and denominator by 1 to obtain the equivalent fraction with a denominator of 16.

$$\frac{3}{16} \cdot \frac{1}{1} = \boxed{\frac{3}{16}}$$

Your Turn Problem #5

Write the following fraction as an equivalent fraction with the given denominator:

$$\frac{11}{38} = \frac{}{38}$$

6.1 Homework: Reviewing the LCM and Equivalent Fractions

Find the LCM for the following

1. 15, 20

2. 8, 12

3. 21, 28

4. 7, 12

5. 5, 11

6. 3, 12

7. 26, 39

8. 34, 51

9. 9, 15

10. 15, 20, 25

11. 21, 15, 35

12. 84, 90

6.1 Homework: Reviewing the LCM and Equivalent Fractions cont.

Write an equivalent fraction with the given denominator.

13. $\dfrac{3}{4} = \dfrac{}{32}$

14. $\dfrac{5}{17} = \dfrac{}{51}$

15. $\dfrac{4}{1} = \dfrac{}{16}$

16. $5 = \dfrac{}{25}$

17. $7 = \dfrac{}{8}$

18. $9 = \dfrac{}{6}$

19. $\dfrac{15}{16} = \dfrac{}{64}$

20. $\dfrac{11}{18} = \dfrac{}{54}$

21. $\dfrac{6}{14} = \dfrac{}{98}$

22. $\dfrac{5}{14} = \dfrac{}{42}$

23. $\dfrac{7}{24} = \dfrac{}{168}$

24. $\dfrac{5}{13} = \dfrac{}{169}$

25. $\dfrac{3}{8} = \dfrac{}{424}$

26. $\dfrac{9}{16} = \dfrac{}{272}$

27. $\dfrac{17}{40} = \dfrac{}{800}$

28. $\dfrac{8}{25} = \dfrac{}{550}$

29. $\dfrac{4}{17} = \dfrac{}{17}$

30. $\dfrac{11}{40} = \dfrac{}{40}$

6.2 Addition with Fractions

Like Denominators: Two or more fractions are said to have *like* denominators when they both contain the same denominator.

For example; $\dfrac{5}{12}$ and $\dfrac{1}{12}$ have like denominators

---

Adding Fractions with Like Denominators

1. Add the Numerators. Write the result over the original denominator.

2. Reduce if possible.

Example 1. Add $\dfrac{5}{24} + \dfrac{7}{24}$

Solution: $\dfrac{5}{24} + \dfrac{7}{24} = \dfrac{5+7}{24} = \dfrac{12}{24}$

Now reduce: $\dfrac{12}{24} = \dfrac{\cancel{2} \cdot \cancel{2} \cdot \cancel{3}}{\cancel{2} \cdot \cancel{2} \cdot 2 \cdot \cancel{3}} = \boxed{\dfrac{1}{2}}$ or $\dfrac{12 \div 12}{24 \div 12} = \boxed{\dfrac{1}{2}}$

Your Turn Problem #1

Add $\dfrac{3}{20} + \dfrac{11}{20}$

Unlike Denominators: Two or more fractions are said to have unlike denominators when they both contain different denominators.

The LCD is the lowest common denominator. The LCD is the LCM for the denominators (same process).

Adding Fractions with Unlike Denominators

1. Find the LCD for each fraction.

2. Write each fraction as an equivalent fraction using the LCD for the denominator.

   (Note: Some students prefer to write the fractions vertically.)

3. Add the numerators. Write the result over the LCD.

4. Reduce if possible.

Example 2. Add $\dfrac{5}{12} + \dfrac{3}{8}$

Solution: First, write the problem vertically and find the LCD (LCM for 12 and 8).

$$\begin{array}{r} \dfrac{5}{12} \\[2mm] +\dfrac{3}{8} \\ \hline \end{array}$$

$$12 = 2 \cdot 2 \cdot 3$$
$$8 = 2 \cdot 2 \cdot 2$$
$$LCD = 2 \cdot 2 \cdot 2 \cdot 3$$
$$= 24$$

Write as equivalent fractions. Add the like fractions and reduce if possible.

$$\begin{array}{r} \dfrac{5}{12} \cdot \dfrac{2}{2} = \dfrac{10}{24} \\[2mm] +\dfrac{3}{8} \cdot \dfrac{3}{3} = \dfrac{9}{24} \\ \hline \end{array}$$

Answer: $\boxed{\dfrac{19}{24}}$

Your Turn Problem #2

Add $\dfrac{5}{18} + \dfrac{2}{15}$

Example 3. Add $\dfrac{4}{35} + \dfrac{8}{21}$

Solution: First, write the problem vertically and find the LCD for 35 and 21 (same as LCM).

$$\begin{array}{r} \dfrac{4}{35} \\[2mm] + \dfrac{8}{21} \\ \hline \end{array}$$

$$\begin{aligned} 35 &= 5\cdot 7 \\ 21 &= 3\cdot 7 \\ LCD &= 3\cdot 5\cdot 7 \\ &= 105 \end{aligned}$$

Write as equivalent fractions. Add the like fractions and reduce if possible.

$$\begin{array}{r} \dfrac{4}{35}\cdot\dfrac{3}{3} = \dfrac{12}{105} \\[3mm] + \dfrac{8}{21}\cdot\dfrac{5}{5} = \dfrac{40}{105} \\ \hline \end{array}$$

Answer: $\boxed{\dfrac{52}{105}}$

Your Turn Problem #3

Add $\dfrac{5}{24} + \dfrac{11}{30}$

Example 4. Find the perimeter of a rectangle if the width is $\dfrac{3}{8}$ ft and the length is $\dfrac{5}{6}$ ft.

$\dfrac{3}{8}$ ft

$\dfrac{5}{6}$ ft.

Solution: To find the perimeter, we need to *add all four sides*. (Note: we could also multiply each side by 2 and then add the results.)

$$\dfrac{3}{8}+\dfrac{3}{8}+\dfrac{5}{6}+\dfrac{5}{6}$$

Since this is addition; rewrite it vertically, find the LCD, change to equivalent fractions and add.

$$\dfrac{3}{8}\cdot\dfrac{3}{3}=\dfrac{9}{24}$$

$$\dfrac{3}{8}\cdot\dfrac{3}{3}=\dfrac{9}{24}$$

$$\dfrac{5}{6}\cdot\dfrac{4}{4}=\dfrac{20}{24}$$

$$+\ \dfrac{5}{6}\cdot\dfrac{4}{4}=\dfrac{20}{24}$$

$$=\dfrac{58}{24}\quad \text{Now, change this answer into a mixed number and reduce.}$$

$$=2\dfrac{10}{24}=2\dfrac{5}{12}\quad \boxed{\text{The perimeter of the rectangle is } 2\dfrac{5}{12}\text{ ft.}}$$

Your Turn Problem #4

Find the perimeter of a triangle if the sides are $\dfrac{5}{12}$ in, $\dfrac{2}{9}$ in, and $\dfrac{4}{15}$ in.

6.2 Homework: Addition with Fractions

Perform the indicated operation. Do not leave improper fractions for answers.

1. $\dfrac{3}{5} + \dfrac{1}{5}$

2. $\dfrac{2}{9} + \dfrac{1}{9}$

3. $\dfrac{11}{15} + \dfrac{4}{15}$

4. $\dfrac{5}{18} + \dfrac{7}{18}$

5. $\dfrac{19}{20} + \dfrac{11}{20}$

6. $\dfrac{3}{4} + \dfrac{1}{4}$

7. $\dfrac{6}{13} + \dfrac{4}{13} + \dfrac{5}{13}$

8. $\dfrac{7}{12} + \dfrac{11}{12} + \dfrac{5}{12}$

9. $\dfrac{2}{3} + \dfrac{1}{4}$

10. $\dfrac{1}{6} + \dfrac{3}{8}$

6.2 Homework: Addition with Fractions cont.

Perform the indicated operation. Do not leave improper fractions for answers.

11. $\dfrac{5}{12}+\dfrac{3}{8}$

12. $\dfrac{5}{18}+\dfrac{7}{12}$

13. $\dfrac{7}{20}+\dfrac{2}{15}$

14. $\dfrac{5}{12}+\dfrac{7}{24}$

15. $\dfrac{6}{25}+\dfrac{3}{10}$

16. $\dfrac{7}{12}+\dfrac{5}{6}$

17. $\dfrac{3}{8}+\dfrac{5}{12}$

18. $\dfrac{5}{24}+\dfrac{9}{40}$

6.2 Homework: Addition with Fractions cont.
Perform the indicated operation. Do not leave improper fractions for answers.

19. $\dfrac{5}{84} + \dfrac{7}{90}$

20. $\dfrac{7}{18} + \dfrac{13}{20}$

21. $\dfrac{15}{22} + \dfrac{8}{33}$

22. $\dfrac{8}{45} + \dfrac{7}{15}$

23. $\dfrac{5}{8} + \dfrac{3}{20}$

24. $\dfrac{11}{18} + \dfrac{1}{24}$

6.2 Homework: Addition with Fractions cont.

Perform the indicated operation. Do not leave improper fractions for answers.

25. $\dfrac{1}{8} + \dfrac{3}{10} + \dfrac{4}{15}$

26. $\dfrac{7}{12} + \dfrac{5}{18} + \dfrac{5}{24}$

27. $\dfrac{2}{15} + \dfrac{8}{25} + \dfrac{22}{63}$

28. $\dfrac{12}{35} + \dfrac{8}{45} + \dfrac{11}{30}$

Solve each application. Do not leave improper fractions for answers.

29. Find the perimeter of a rectangle if each width is 5/9 yd and each length is 3/4 yd.

6.2 Homework: Addition with Fractions cont.

Solve each application. Do not leave improper fractions for answers.

30. A countertop consists of a board 3/4 inches (in) thick and tile 3/8 in thick. What is the overall thickness?

31. Leah budgets 2/5 of her income for housing and 1/6 of her income for food. What fraction of her income is budgeted for these two purposes?

32. Find the perimeter of a triangle if the sides are 4/15 ft, 3/10 ft, and 7/24 ft.

6.3 Subtracting and Comparing Fractions

Subtracting Fractions with Like Denominators

1. Subtract the numerators. Write the result over the original denominator.

2. Reduce if possible.

Example 1. Subtract: $\dfrac{17}{24} - \dfrac{7}{24}$

Solution: $\dfrac{17}{24} - \dfrac{7}{24} = \dfrac{17-7}{24} = \dfrac{10}{24}$

Now reduce, $\dfrac{10}{24} = \dfrac{\cancel{2}\cdot 5}{\cancel{2}\cdot 2\cdot 2\cdot 3} = \boxed{\dfrac{5}{12}}$

Your Turn Problem #1

Subtract: $\dfrac{11}{20} - \dfrac{1}{20}$

Unlike Denominators: Two or more fractions are said to have unlike denominators when they both contain different denominators.

The LCD is the lowest common denominator. The LCD is the LCM for the denominators (same process).

Subtracting Fractions with Unlike Denominators

1. Find the LCD for each fraction.

2. Write each fraction as an equivalent fraction using the LCD for the denominator.

   (Note: Some students prefer to write the fractions vertically.)

3. Subtract the Numerators. Write the result over the original denominator.

4. Reduce if possible.

Example 2. Subtract: $\dfrac{5}{12} - \dfrac{3}{8}$

Solution: First, write the problem vertically and find the LCD (LCM for 12 and 8)

$$\begin{array}{r} \dfrac{5}{12} \\[2mm] -\dfrac{3}{8} \\ \hline \end{array}$$

$$\begin{aligned} 12 &= 2 \cdot 2 \cdot 3 \\ 8 &= 2 \cdot 2 \cdot 2 \\ LCD &= 2 \cdot 2 \cdot 2 \cdot 3 \\ &= 24 \end{aligned}$$

Write as equivalent fractions. Subtract the like fractions and reduce if possible.

$$\begin{array}{r} \dfrac{5}{12} \cdot \dfrac{2}{2} = \dfrac{10}{24} \\[3mm] -\dfrac{3}{8} \cdot \dfrac{3}{3} = \dfrac{9}{24} \\ \hline \end{array}$$

Answer: $\boxed{\dfrac{1}{24}}$

Your Turn Problem #2

Subtract: $\dfrac{5}{18} - \dfrac{2}{15}$

Example 3. Subtract: $\dfrac{19}{35} - \dfrac{8}{21}$

Solution: First, write the problem vertically and find the LCD for 35 and 21 (same as LCM).

$$\begin{array}{r} \dfrac{19}{35} \\[2mm] -\dfrac{8}{21} \\ \hline \end{array}$$

$$\begin{aligned} 35 &= 5 \cdot 7 \\ 21 &= 3 \cdot 7 \\ LCD &= 3 \cdot 5 \cdot 7 \\ &= 105 \end{aligned}$$

Write as equivalent fractions. Add the like fractions and reduce if possible.

$$\begin{array}{r} \dfrac{19}{35} \cdot \dfrac{3}{3} = \dfrac{57}{105} \\[3mm] -\dfrac{8}{21} \cdot \dfrac{5}{5} = \dfrac{40}{105} \\ \hline \end{array}$$

Answer: $\boxed{\dfrac{17}{105}}$

Your Turn Problem #3

Subtract: $\dfrac{13}{24} - \dfrac{7}{30}$

Example 4. Perform the indicated operation: $\dfrac{7}{15} - \dfrac{3}{10} + \dfrac{5}{12}$

Solution: Find the LCD. Write as equivalent fractions. Then, according to the Order of Operations, work left to right with addition and subtraction.

$$15 = 3 \cdot 5$$
$$10 = 2 \cdot 5$$
$$12 = 2 \cdot 2 \cdot 3$$
$$LCD = 2 \cdot 2 \cdot 3 \cdot 5$$
$$= 60$$

$$\dfrac{7}{15} \cdot \dfrac{4}{4} - \dfrac{3}{10} \cdot \dfrac{6}{6} + \dfrac{5}{12} \cdot \dfrac{5}{5}$$

$$\dfrac{28}{60} - \dfrac{18}{60} + \dfrac{25}{60} = \dfrac{28 - 18 + 25}{60} = \dfrac{35 \div 5}{60 \div 5}$$

$$\boxed{\dfrac{7}{12}}$$

Answer:

Your Turn Problem #4

Perform the indicated operation. Do not leave an improper fraction for an answer.

$$\dfrac{11}{15} - \dfrac{1}{6} + \dfrac{5}{9}$$

Comparing Fractions

Looking at the figure below, we can see that $\dfrac{5}{8}$ is larger than $\dfrac{3}{8}$.

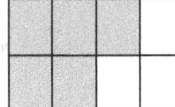 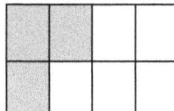

We can use the inequality symbols ($>$ and $<$) to show which number is greater.

    $a > b$; a is greater than b $a < b$; a is less than b

For our example above, $\dfrac{5}{8}$ is greater than $\dfrac{3}{8}$, we can write $\dfrac{5}{8} > \dfrac{3}{8}$.

To determine which of two fractions is greater.

If the denominators are common (the same), compare the numerators.

If the denominators are different: Change each fraction to an equivalent fraction using the LCM

for the denominator. Then compare the numerators to determine which is larger.

Example 5. Use $>$ or $<$ between the two fractions to write a true sentence.

$$\frac{7}{15} \qquad \frac{11}{24}$$

Solution: Change each fraction to an equivalent fraction using the LCM as the denominator.

$$\frac{7}{15} \cdot \frac{8}{8} = \frac{56}{120} \qquad \frac{11}{24} \cdot \frac{5}{5} = \frac{55}{120}$$

We can now determine the fraction on the left is larger, $56 > 55$.

Answer: $\boxed{\dfrac{7}{15} > \dfrac{11}{24}}$

Your Turn Problem #5

Use $>$ or $<$ between the two fractions to write a true sentence: $\dfrac{7}{18} \qquad \dfrac{9}{20}$

Ordering Fractions

If the denominators of two or more fractions are identical, then we can arrange the fractions by simply looking at the numerators.

Example: Arrange from smallest to largest.

$$\frac{4}{11}, \frac{7}{11}, \frac{2}{11} \qquad \textbf{Answer:} \quad \frac{2}{11}, \frac{4}{11}, \frac{7}{11}$$

If the denominators are not the same, write each as an equivalent fraction using the LCD.

Example 6: Arrange from smallest to largest: $\dfrac{5}{9}, \dfrac{7}{12}, \dfrac{13}{27}$

Solution: Find the LCD. Write each as an equivalent fraction using the LCD.

$$\begin{aligned}
9 &= 3 \cdot 3 \\
12 &= 2 \cdot 2 \cdot 3 \\
27 &= 3 \cdot 3 \cdot 3 \\
LCD &= 2 \cdot 2 \cdot 3 \cdot 3 \cdot 3 \\
&= 108
\end{aligned}
\qquad
\frac{5}{9} \cdot \frac{12}{12} = \frac{60}{108}, \quad \frac{7}{12} \cdot \frac{9}{9} = \frac{63}{108}, \quad \frac{13}{27} \cdot \frac{4}{4} = \frac{52}{108}$$

Now that the denominators are identical, we can determine the larger fraction. The third fraction is the smallest and the middle fraction is the largest.

Answer: $\boxed{\dfrac{13}{27}, \dfrac{5}{9}, \dfrac{7}{12}}$

Your Turn Problem #6

Arrange from smallest to largest.

$$\frac{11}{18}, \frac{7}{9}, \frac{2}{3}$$

6.3 Homework: Subtracting and Comparing Fractions

Perform the indicated operation. Do not leave improper fractions for answers.

1. $\dfrac{18}{25} - \dfrac{8}{25}$

2. $\dfrac{7}{15} - \dfrac{4}{15}$

3. $\dfrac{8}{10} - \dfrac{7}{10} + \dfrac{3}{10}$

4. $\dfrac{5}{24} - \dfrac{1}{24} + \dfrac{5}{24}$

5. $\dfrac{7}{9} - \dfrac{1}{4}$

6. $\dfrac{11}{16} - \dfrac{5}{12}$

7. $\dfrac{7}{8} - \dfrac{5}{12}$

8. $\dfrac{11}{24} - \dfrac{9}{40}$

6.3 Homework: Subtracting and Comparing Fractions cont.

Perform the indicated operation. Do not leave improper fractions for answers.

9. $\dfrac{25}{84} - \dfrac{7}{90}$

10. $\dfrac{13}{18} - \dfrac{7}{20}$

11. $\dfrac{15}{22} - \dfrac{8}{33}$

12. $\dfrac{41}{45} - \dfrac{7}{15}$

13. $\dfrac{5}{8} - \dfrac{3}{20}$

14. $\dfrac{11}{18} - \dfrac{1}{24}$

6.3 Homework: Subtracting and Comparing Fractions cont.

Perform the indicated operation. Do not leave improper fractions for answers.

15. $\dfrac{7}{8} - \dfrac{3}{10} - \dfrac{4}{15}$

16. $\dfrac{11}{12} - \dfrac{5}{18} - \dfrac{5}{24}$

17. $\dfrac{14}{15} - \dfrac{8}{25} - \dfrac{22}{63}$

18. $\dfrac{34}{35} - \dfrac{8}{45} - \dfrac{11}{30}$

19. $\dfrac{11}{15} - \dfrac{3}{20} + \dfrac{7}{45}$

20. $\dfrac{23}{48} - \dfrac{17}{80} + \dfrac{11}{30}$

6.3 Homework: Subtracting and Comparing Fractions cont.
Solve the following applications.

21. Maria, a road inspector, must inspect 7/8 of a mile of road. If she has already inspected 1/3 of a mile, how much more does she need to inspect?

22. In Redlands, 3/14 of the adult residents have college degrees while 7/12 of the adult residents have high school degrees only. What fraction represents the difference of high school degree resident's and college degree residents?

23. In Yucaipa, California, 4/9 of the eligible voters are Democrat, and 7/15 are Republican. What fraction represents how many more voters are Republican than Democrat?

6.3 Homework: Subtracting and Comparing Fractions cont.
Solve the following applications.

24. A hamburger that weighed $\frac{1}{4}$ pound before cooking was $\frac{3}{16}$ of a pound after cooking. How much weight was lost during cooking?

25. A container holds $\frac{7}{8}$ of a gallon of a liquid. If $\frac{1}{3}$ of a gallon is drained from the container, how much of the liquid remains in the container?

26. John lives 3/8 of a mile from the Museum of Science. Sylvia lives 1/4 of a mile from the Museum of Science. How much closer is Sylvia from the museum?

27. As part of a unit on earthquakes, a science class is researching the movement of a nearby fault line. The fault line moved 9/10 of an inch this year and 1/2 of an inch the year before. How much more did the fault line move this year than it did last year?

6.3 Homework: Subtracting and Comparing Fractions cont.

Complete the statements, using the symbol < or >.

28. $\dfrac{3}{4}$ $\dfrac{7}{10}$

29. $\dfrac{5}{16}$ $\dfrac{5}{12}$

30. $\dfrac{5}{6}$ $\dfrac{11}{15}$

31. $\dfrac{17}{24}$ $\dfrac{13}{36}$

32. $\dfrac{7}{9}$ $\dfrac{3}{4}$

33. $\dfrac{4}{7}$ $\dfrac{6}{11}$

6.3 Homework: Subtracting and Comparing Fractions cont.

Arrange the given fractions from smallest to largest

34. $\dfrac{7}{20}$, $\dfrac{5}{12}$, $\dfrac{3}{8}$

35. $\dfrac{2}{3}$, $\dfrac{9}{16}$, $\dfrac{7}{12}$

36. $\dfrac{17}{45}$, $\dfrac{12}{25}$, $\dfrac{13}{30}$

37. $\dfrac{9}{40}$, $\dfrac{1}{5}$, $\dfrac{5}{24}$

6.4 Addition with Mixed Numbers

A mixed number is the sum of a *whole number* and a *proper fraction*.

Examples: $4\dfrac{7}{18}$, $77\dfrac{3}{5}$

If a number is in the form of a mixed number, however the fraction part is an improper fraction:

If the fractional part of a mixed number is improper:

1. Change the improper fraction to a mixed number.

2. Add *both* the whole numbers.

Example 1. Write as a mixed number: $15\dfrac{9}{4}$

Solution: Since $\dfrac{9}{4}$ is improper, change it to a mixed number.

$$4\overline{)9} = 2\dfrac{1}{4}$$

$$15\dfrac{9}{4} = 15 + 2\dfrac{1}{4} = \boxed{17\dfrac{1}{4}}$$

Your Turn Problem #1

Write as a mixed number: $7\dfrac{11}{9}$

Procedure: Adding Mixed Numbers

1. Rewrite the problem vertically aligning the whole numbers and the fractions.

2. Add the fractions. If the result is an improper fraction, convert it to a mixed number.

3. Add all of the whole numbers.

Example 2. Add $78\dfrac{5}{12} + 37\dfrac{3}{8}$

Solution: First, write the problem vertically and find the LCD (LCM for 12 and 8).

$$
\begin{array}{r}
78\dfrac{5}{12} \\[2mm]
+\ 37\dfrac{3}{8} \\[2mm]
\hline
\end{array}
\qquad
\begin{aligned}
12 &= 2\cdot 2\cdot 3 \\
8 &= 2\cdot 2\cdot 2 \\
\text{LCD} &= 2\cdot 2\cdot 2\cdot 3 \\
&= 24
\end{aligned}
$$

Write as equivalent fractions. Add the like fractions and reduce if possible. Then add the whole numbers.

$$
\begin{array}{r}
78\dfrac{5}{12}\cdot\dfrac{2}{2} = 78\dfrac{10}{24} \\[3mm]
+\ 37\dfrac{3}{8}\cdot\dfrac{3}{3} =\ 37\dfrac{9}{24} \\[2mm]
\hline
115\ \dfrac{19}{24} \quad \leftarrow \text{proper fraction}
\end{array}
$$

Answer: $\boxed{115\dfrac{19}{24}}$

Your Turn Problem #2

Add $95\dfrac{1}{4} + 56\dfrac{3}{10}$

$$95\dfrac{1}{4} \cdot \dfrac{\quad}{\quad} = 95\dfrac{\quad}{\quad}$$

$$+\ 56\dfrac{3}{10} \cdot \dfrac{\quad}{\quad} = 56\dfrac{\quad}{\quad}$$

Example 3. Add $84\dfrac{7}{9} + 43\dfrac{5}{6}$

Solution: First, write the problem vertically and find the LCD (LCM for 6 and 9)

$$\begin{array}{r} 84\dfrac{7}{9} \\[2mm] +\ 43\dfrac{5}{6} \\ \hline \end{array} \qquad \begin{array}{l} 6 = 2 \cdot 3 \\ 9 = 3 \cdot 3 \\ \text{LCD} = 2 \cdot 3 \cdot 3 \\ \qquad = 18 \end{array}$$

Write as equivalent fractions. Add the like fractions and reduce if possible. Then add the whole numbers.

$$\begin{array}{r} 84\dfrac{7}{9} \cdot \dfrac{2}{2} = 84\dfrac{14}{18} \\[3mm] +\ 43\dfrac{5}{6} \cdot \dfrac{3}{3} = 37\dfrac{15}{18} \\ \hline 127\dfrac{29}{18} \end{array} \quad \leftarrow \text{Since the sum is an improper fraction, change it to a mixed number.}$$

$$127\ +\ 1\dfrac{11}{18} \qquad \textbf{Answer:}\ \boxed{128\dfrac{11}{18}}$$

Note: The sum of the fractional parts in this problem was more than 1. After changing the improper fraction to a mixed number, we "carried over" the 1 to the whole numbers.

Your Turn Problem #3

Add $210\dfrac{13}{20}+135\dfrac{11}{15}$

$210\dfrac{13}{20}\cdot\text{——}=210\text{——}$

$+\ 135\dfrac{11}{15}\cdot\text{——}=135\text{——}$

Example 4.   Add $42+11\dfrac{4}{13}$

Solution: $\begin{array}{r}42\\+\ 11\dfrac{4}{13}\\\hline\end{array}$

No LCD necessary since there is only one fraction. Add whole numbers and bring down the fraction.

Answer: $\boxed{53\dfrac{4}{13}}$

Your Turn Problem #4

Add $57\dfrac{11}{14}+28$

Example 5. Add $11\dfrac{1}{2}+\dfrac{2}{3}$

Solution: First, write the problem vertically and find the LCD (LCM for 2 and 3).

$$11\dfrac{1}{2}$$
$$+\ \dfrac{2}{3}$$

$$2 = 2$$
$$3 = 3$$
$$\text{LCD} = 2 \cdot 3$$
$$= 6$$

Write as equivalent fractions. Add the like fractions and reduce if possible. Then add the whole numbers.

$$11\dfrac{1}{2} \cdot \dfrac{3}{3} = 11\dfrac{3}{6}$$

$$+\quad \dfrac{2}{3} \cdot \dfrac{2}{2} = \dfrac{4}{6}$$

$$11\dfrac{7}{6} \quad \leftarrow \text{Since the sum is an improper fraction, change it to a mixed number.}$$

$$11\ +\ 1\dfrac{1}{6}$$

Answer: $\boxed{12\dfrac{1}{6}}$

Your Turn Problem #5

Add $23\dfrac{3}{4}+\dfrac{4}{5}$

Example 6. Find the perimeter of a rectangle if the width is $2\frac{1}{4}$ ft and the length is $5\frac{2}{3}$ ft.

$2\frac{1}{4}$ ft $\boxed{}$

$5\frac{2}{3}$ ft.

Solution: To find the perimeter, we need to add all *four* sides. We could write all four mixed numbers vertically and add all four mixed numbers at the same time. We could also add the two widths first, then the two lengths, then add the results. Each method will give the same answer.

$$\underbrace{2\frac{1}{4}+2\frac{1}{4}}+\underbrace{5\frac{2}{3}+5\frac{2}{3}}$$

$$
\begin{array}{c}
2\frac{1}{4} \\
+\ 2\frac{1}{4} \\
\hline
4\frac{2}{4}\ =4\frac{1}{2}
\end{array}
\qquad\qquad
\begin{array}{c}
5\frac{2}{3} \\
+\ 5\frac{2}{3} \\
\hline
10\frac{4}{3}\ =10+1\frac{1}{3}=11\frac{1}{3}
\end{array}
$$

Now, add the two results.

$$= 4\frac{1}{2}+11\frac{1}{3} \;\rightarrow\;
\begin{array}{c}
4\frac{1}{2}\cdot\frac{3}{3}\ =\ 4\frac{3}{6} \\
+\ 11\frac{1}{3}\cdot\frac{2}{2}=11\frac{2}{6} \\
\hline
15\frac{5}{6}
\end{array}
$$

Answer: $\boxed{\text{The perimeter of the rectangle is } 15\frac{5}{6} \text{ ft.}}$

Your Turn Problem #6

Find the perimeter of a rectangle if the length is $12\frac{5}{6}$ ft and the width is $3\frac{5}{8}$ ft.

6.4 Homework: Addition with Mixed Numbers

Perform the indicated operation. Do not leave improper fractions for answers.

1. $12\dfrac{3}{8}+5\dfrac{1}{8}$

2. $17\dfrac{5}{16}+11\dfrac{3}{16}$

3. $6\dfrac{2}{9}+5\dfrac{4}{9}$

4. $72\dfrac{7}{20}+15\dfrac{3}{20}$

5. $5\dfrac{3}{8}+4\dfrac{1}{6}$

6. $12\dfrac{3}{14}+5\dfrac{2}{7}$

6.4 Homework: Addition with Mixed Numbers

Perform the indicated operation. Do not leave improper fractions for answers.

7. $11\frac{2}{9}+4\frac{5}{12}$

8. $9\frac{5}{6}+2\frac{3}{4}$

9. $15\frac{3}{5}+4\frac{7}{10}$

10. $9\frac{11}{16}+8\frac{7}{12}$

11. $23\frac{5}{6}+13\frac{7}{9}$

12. $263\frac{11}{15}+178\frac{7}{10}$

6.4 Homework: Addition with Mixed Numbers

Perform the indicated operation. Do not leave improper fractions for answers.

13. $61\dfrac{7}{11}+41\dfrac{1}{2}$

14. $22\dfrac{7}{8}+15\dfrac{5}{9}$

15. $54\dfrac{7}{12}+36\dfrac{14}{15}$

16. $14\dfrac{13}{24}+7\dfrac{15}{16}$

17. $5\dfrac{9}{10}+\dfrac{3}{4}$

18. $15\dfrac{8}{15}+3\dfrac{11}{20}$

6.4 Homework: Addition with Mixed Numbers

Perform the indicated operation. Do not leave improper fractions for answers.

19. $12 + 5\dfrac{7}{8}$

20. $15\dfrac{1}{2} + 3\dfrac{2}{3} + 8\dfrac{3}{4}$

21. $17\dfrac{7}{9} + \dfrac{2}{3}$

22. $18\dfrac{5}{12} + 19\dfrac{1}{3} + 20\dfrac{1}{4}$

Solve the following applications. Do not leave improper fractions for answers.

23. A plumber needs pieces of pipe 15 ½ ft and 25 ¾ ft long. What is the total length of pipe that is needed?

6.4 Homework: Addition with Mixed Numbers

Solve the following applications. Do not leave improper fractions for answers.

24. In Seattle, they received 4 2/7 inches of rain in December, 3 5/6 inches of rain in January, and 5 8/21 inches of rain in February. What is the total amount of rainfall for the three months?

25. Find the perimeter of a rectangle if the with is $15\dfrac{1}{3}$ ft and the length is $28\dfrac{3}{4}$ ft.

26. Find the perimeter of the triangle.

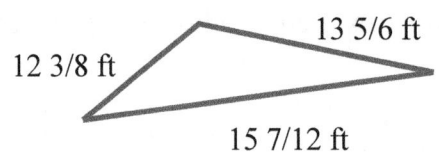

12 3/8 ft

13 5/6 ft

15 7/12 ft

6.5 Subtraction with Mixed Numbers

When performing subtraction, if the bottom digit is larger than the top digit, it is necessary to borrow.

We borrow from the left and write the 1 in front of the number on its right.

Example:

$$\begin{array}{r} 72 \\ -\ 47 \\ \hline \end{array}$$

Solution:

$$\begin{array}{r} \overset{6}{\cancel{7}}\ ^{1}2 \\ -\ 4\ \ 7 \\ \hline 2\ \ 5 \end{array}$$

Answer: $\boxed{25}$

In this section of Subtraction with Mixed Numbers, the process of borrowing will again be necessary to borrow when the bottom fraction is larger than the top fraction.

Note: Any number over itself equals 1. Examples: $\dfrac{12}{12}=1,\ \dfrac{15}{15}=1$

So if we add 1 to a fraction, the 1 can be written as any number over itself. The number chosen will be the same as the denominator of the fraction the 1 is being added to.

Example 1. Add $\dfrac{7}{30}+1$ (Do not write as a mixed number.)

Solution: We can change the 1 to any number over itself. Since the denominator of the first fraction is 30, we should rewrite the 1 as $\dfrac{30}{30}$.

$$\frac{7}{30}+1 \ \Rightarrow\ \frac{7}{30}+\frac{30}{30}$$

Answer: $\boxed{\dfrac{37}{30}}$

Alternative method:

$$\frac{7}{30}+1 \ \Rightarrow\ 1\frac{7}{30} \ \ \text{Now, change the mixed number to an improper fraction.}$$

Answer: $\boxed{\dfrac{37}{30}}$

Your Turn Problem #1

Add $\dfrac{5}{18} + 1$ Do not write as a mixed number.

Procedure: Subtracting Mixed Numbers

1. Rewrite the problem vertically aligning the whole numbers and the fractions.

2. Subtract the fractions. If the top fraction is smaller than the bottom fraction, borrow a 1 from the top whole number. Add the 1 to the top fraction.

3. Subtract the fractions and the whole numbers

Example 2. Subtract: $37\dfrac{7}{8} - 25\dfrac{2}{3}$

Solution: First, write the problem vertically and find the LCD (LCM for 8 and 3). Write as equivalent fractions and subtract.

$$
\begin{aligned}
&37\dfrac{7}{8} \\[2mm]
-\;&25\dfrac{2}{3}
\end{aligned}
\qquad\qquad
\begin{aligned}
8 &= 2\cdot 2\cdot 2 \\
3 &= 3 \\
LCD &= 2\cdot 2\cdot 2\cdot 3 \\
&= 24
\end{aligned}
$$

$$
\begin{aligned}
37\dfrac{7}{8}\cdot\dfrac{\;\;}{\;\;} &= 37\dfrac{\;\;}{24} \\[2mm]
-\;25\dfrac{2}{3}\cdot\dfrac{\;\;}{\;\;} &= 25\dfrac{\;\;}{24}
\end{aligned}
\;\;\rightarrow\;\;
\begin{aligned}
37\dfrac{7}{8}\cdot\dfrac{3}{3} &= 37\dfrac{21}{24} \\[2mm]
-\;25\dfrac{2}{3}\cdot\dfrac{8}{8} &= 25\dfrac{16}{24} \\
\hline
&\;\;12\;\dfrac{5}{24}
\end{aligned}
$$

Since the top fraction is larger, there is no need to borrow. We can subtract.

Answer: $\boxed{12\dfrac{5}{24}}$

Your Turn Problem #2

Subtract: $46\dfrac{7}{12} - 12\dfrac{2}{5}$

$$46\dfrac{7}{12} \cdot \dfrac{\ \ }{\ \ } = 46\dfrac{\ \ }{\ \ }$$

$$-\ 12\dfrac{2}{5} \cdot \dfrac{\ \ }{\ \ } = 12\dfrac{\ \ }{\ \ }$$

Example 3. Subtract: $28\dfrac{1}{3} - 13\dfrac{4}{5}$

Solution: First, write the problem vertically and find the LCD (LCM for 3 and 5). Write as equivalent fractions and subtract if possible.

$$28\dfrac{1}{3} \cdot \dfrac{5}{5} = 28\dfrac{5}{15}$$
$$-\ 13\dfrac{4}{5} \cdot \dfrac{3}{3} = 13\dfrac{12}{15}$$

Since the bottom fraction is larger, borrow a 1 from the top whole number in the form of $\dfrac{15}{15}$. Add it to the top fraction. Then we can subtract.

The *alternative method* is to write the "1" in front of the fraction and convert the mixed number to an improper fraction. Then subtract.

Alternative method

$$28\dfrac{1}{3} \cdot \dfrac{5}{5} = \overset{27}{\cancel{28}}\dfrac{5}{15} + \dfrac{15}{15} \qquad 27\dfrac{20}{15}$$
$$-\ 13\dfrac{4}{5} \cdot \dfrac{3}{3} = 13\dfrac{12}{15} \qquad\qquad -\ 13\dfrac{12}{15}$$

$\rightarrow$

$$28\dfrac{1}{3} \cdot \dfrac{5}{5} = \overset{27}{\cancel{28}}\dfrac{5}{15} = 27 + 1\dfrac{5}{15} = 27\dfrac{20}{15}$$
$$-\ 13\dfrac{4}{5} \cdot \dfrac{3}{3} = 13\dfrac{12}{15} = 13\dfrac{12}{15} = 13\dfrac{12}{15}$$

Answer: $\boxed{14\dfrac{8}{15}}$ $\qquad\qquad$ $\boxed{14\dfrac{8}{15}}$

Your Turn Problem #3

Subtract: $71\dfrac{1}{4} - 18\dfrac{2}{3}$

Example 4. Subtract: $84\dfrac{2}{9} - 51\dfrac{5}{6}$

Solution: First, write the problem vertically and find the LCD (LCM for 9 and 6). Write as equivalent fractions and subtract if possible.

$$84\dfrac{2}{9}\cdot\dfrac{2}{2} = 84\dfrac{4}{18}$$

$$-\ 51\dfrac{5}{6}\cdot\dfrac{3}{3} = 51\dfrac{15}{18}$$

Since the bottom fraction is larger, borrow a 1 from the top whole number in the form of $\dfrac{18}{18}$. Add it to the top fraction. Then we can subtract.

Alternative method

$$84\dfrac{2}{9}\cdot\dfrac{2}{2} = \overset{83}{\cancel{84}}\dfrac{4}{18} + \dfrac{18}{18} \qquad 83\dfrac{22}{18}$$

$$-\ 51\dfrac{5}{6}\cdot\dfrac{3}{3} = 51\dfrac{15}{18} \qquad\qquad -\ 51\dfrac{15}{18}$$

$$\boxed{32\dfrac{7}{18}}$$

$$84\dfrac{2}{9}\cdot\dfrac{2}{2} = \overset{83}{\cancel{84}}\dfrac{4}{18} = 83 + 1\dfrac{4}{18} = 83\dfrac{22}{18}$$

$$-\ 51\dfrac{5}{6}\cdot\dfrac{3}{3} = 51\dfrac{15}{18} = \quad 51\dfrac{15}{18} = 51\dfrac{15}{18}$$

Answer:

$$\boxed{32\dfrac{7}{18}}$$

Your Turn Problem #4

Subtract: $29\dfrac{3}{20} - 12\dfrac{11}{15}$

Example 5. Subtract: $75\dfrac{3}{14} - 25$

Solution:
$$
\begin{array}{r}
75\dfrac{3}{14} \\
-\ 25 \\
\hline
\end{array}
$$

Since the top fraction is larger, there is no need to borrow. We can subtract (bring down the top fraction). Any number minus 0 is the same number.

Answer: $\boxed{50\dfrac{3}{14}}$

Your Turn Problem #5

Subtract: $39\dfrac{8}{15} - 19$

Example 6. Subtract: $68 - 20\dfrac{2}{9}$

Solution:

$$\begin{array}{r} 68 \\ -\ 20\dfrac{2}{9} \\ \hline \end{array}$$

There is no top fraction. Borrow 1 from the top whole number.

Since the bottom fraction has a denominator of 9, the 1 we borrow

will be in the form of $\dfrac{9}{9}$.

$$\begin{array}{r} \overset{67}{\cancel{68}}\quad \dfrac{9}{9} \\ -\ 20\quad \dfrac{2}{9} \\ \hline \end{array}$$

Answer: $\boxed{47\dfrac{7}{9}}$

Your Turn Problem #6

Subtract: $40 - 18\dfrac{3}{8}$

Recall Rules for Order of Operations

When addition and subtraction are combined, order of operations dictates we work left to right.

Example 7. Subtract: $12\frac{2}{5} - 8\frac{1}{10} + 4\frac{1}{3}$

Solution:

$$12\frac{2}{5}\cdot\frac{2}{2} = \quad 12\frac{4}{10}$$
$$-\ 8\frac{1}{10}\cdot\frac{1}{1} = \quad -\ 8\frac{1}{10}$$
$$\overline{\qquad\qquad\qquad 4\frac{3}{10}}$$

$$\Rightarrow\ 4\frac{3}{10} + 4\frac{1}{3}$$

$$4\frac{3}{10}\cdot\frac{3}{3} = 4\frac{9}{30}$$
$$+\ 4\frac{1}{3}\cdot\frac{10}{10} = 4\frac{10}{30}$$
$$\overline{\qquad\qquad\qquad 8\frac{19}{30}}$$

Answer: $\boxed{8\frac{19}{30}}$

Your Turn Problem #7

Subtract: $20\frac{1}{4} - 15\frac{2}{3} + 8\frac{5}{6}$

Recall Rules for Order of Operations

When parentheses are present, order of operations dictates any operations within parentheses are to be performed first.

Example 8. Simplify: $20 - \left(5\frac{3}{8} + 3\frac{1}{6}\right)$

Solution: First, perform the addition inside the parentheses.

$$5\frac{3}{8} \cdot \frac{3}{3} = 5\frac{9}{24}$$

$$+ \quad 3\frac{1}{6} \cdot \frac{4}{4} = 3\frac{4}{24}$$

$$\overline{\phantom{+ \quad 3\frac{1}{6} \cdot \frac{4}{4} = }8\frac{13}{24}}$$

Now we can perform the subtraction.

$$20 - 8\frac{13}{24}$$

Solution:
$$\begin{array}{r} 20 \\ - \quad 8\frac{13}{24} \\ \hline \end{array}$$

There is no top fraction. Borrow 1 from the top whole number.

Since the bottom fraction has a denominator of 24, the 1 we borrow will be in the form of $\frac{24}{24}$.

$$\begin{array}{r} \overset{19}{\cancel{20}} \ \frac{24}{24} \\ - \quad 8 \ \frac{13}{24} \\ \hline \end{array}$$

Answer: $\boxed{11\frac{11}{24}}$

Your Turn Problem #8

Simplify: $30 - \left(12\frac{3}{4} + 5\frac{2}{3}\right)$

Example 9. A plumber cuts a piece of $3\frac{5}{8}$ ft piece of pipe from a 12 foot piece of pipe. How much of the pipe remains?

Solution: "How much remains?" This phrase indicates subtraction.

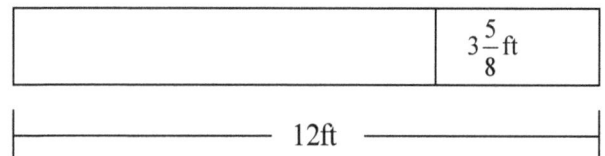

Leftover = Total minus piece cut off.

$$12 - 3\frac{5}{8} \qquad \rightarrow \qquad \begin{array}{r} 12 \\ -\ 3\frac{5}{8} \\ \hline \end{array}$$

There is no top fraction. Borrow 1 from the top whole number.

Since the bottom fraction has a denominator of 8, the 1 we borrow will be in the form of $\frac{8}{8}$.

$$\begin{array}{r} \overset{11}{\cancel{12}}\ \frac{8}{8} \\ -\ 3\ \frac{5}{8} \\ \hline 8\ \frac{3}{8} \end{array}$$

Answer: $\boxed{8\frac{3}{8} \text{ feet of pipe will remain.}}$

Your Turn Problem #9

A newborn typically sleeps an average of $18\frac{7}{15}$ hours each day. How many hours per day is the infant awake? (Hint: There are 24 hours in a day.)

6.5 Homework: Subtraction with Mixed Numbers

Perform the indicated operation. Do not leave improper fractions for answers.

1. $5\dfrac{5}{12} - 2\dfrac{1}{12}$

2. $12\dfrac{3}{4} - 7\dfrac{2}{3}$

3. $20\dfrac{5}{6} - 12\dfrac{2}{9}$

4. $15\dfrac{1}{6} - 6\dfrac{3}{4}$

5. $5\dfrac{1}{3} - 2\dfrac{1}{2}$

6. $15\dfrac{4}{15} - 3\dfrac{11}{20}$

6.5 Homework: Subtraction with Mixed Numbers

Perform the indicated operation. Do not leave improper fractions for answers.

7. $73\dfrac{5}{12} - 67\dfrac{15}{16}$

8. $98\dfrac{2}{9} - 92\dfrac{5}{6}$

9. $20\dfrac{1}{4} - 12\dfrac{2}{3}$

10. $14\dfrac{13}{24} - 7$

11. $5\dfrac{7}{10} - \dfrac{3}{4}$

12. $15 - 3\dfrac{11}{20}$

6.5 Homework: Subtraction with Mixed Numbers cont.

Perform the indicated operation. Do not leave improper fractions for answers.

13. $12 - 5\dfrac{7}{8}$

14. $4 - 2\dfrac{5}{6}$

15. $5 - \dfrac{1}{4}$

16. $6 - \dfrac{2}{5}$

17. $5 - 1\dfrac{4}{9}$

18. $12\dfrac{4}{5} + 5\dfrac{3}{8} - 2\dfrac{7}{10}$

6.5 Homework: Subtraction with Mixed Numbers

Perform the indicated operation. Do not leave improper fractions for answers.

19. $4\dfrac{2}{9} - 2\dfrac{5}{6} + 3\dfrac{1}{2}$

20. $12\dfrac{1}{3} - 5\dfrac{3}{4} + 2\dfrac{1}{6}$

21. $15\dfrac{9}{10} - \left(3\dfrac{1}{5} + 2\dfrac{1}{2}\right)$

22. $28\dfrac{1}{12} - \left(8\dfrac{5}{9} + 4\dfrac{5}{6}\right)$

6.5 Homework: Subtraction with Mixed Numbers cont.

Solve the following applications. Do not leave improper fractions for answers.

23. A roast weighed 5 1/3 lb before cooking and 4 3/4 lb after cooking. How many pounds were lost during the cooking?

24. A roll of paper contains 30 yards. If 16 5/6 yards is cut from the roll, how much paper remains?

25. A contactor needs to cut two pieces of cable from a 50 foot roll of cable. If one of the pieces is $11\frac{3}{4}$ ft long and the other is $12\frac{1}{2}$ ft long, how much of the cable will be left?

6.5 Homework: Subtraction with Mixed Numbers cont.

Solve the following applications. Do not leave improper fractions for answers.

26. The average college student spends 6 3/4 hours sleeping, 4 1/6 hours working at their job, and 5 3/8 hours studying. How many hours in a day does a student have free? (Hint: there are 24 hours in a day.)

27. Kathie's goal is to lose 10 pounds in three months. She lost $3\frac{3}{4}$ lbs during the first month and $3\frac{2}{3}$ lbs the second month. How many more pounds does she have left to lose in the last month to achieve her goal of losing 10 pounds in three months?

6.6 Order of Operations with Fractions

Recall the Order of Operations Agreement.

Procedure: For Order of Operations

Step 1: <u>P</u>arentheses: perform operations inside parentheses or other grouping symbols such as brackets or braces.

Step 2: <u>E</u>xponents: simplify (evaluate) exponential notation expressions

Step 3: <u>M</u>ultiply or <u>d</u>ivide as they appear from left to right

Step 4: <u>A</u>dd or <u>s</u>ubtract as they appear from left to right

---

When you are asked to "simplify" remember to use the Order of Operations Agreement (PEMDAS). In simplifying, work vertical; for each step recopy the entire problem. For each line of the problem, perform only the step being performed in the Order of Operations.

Restating the Rules:

1. Always simplify what is inside the parentheses or grouping symbols first.

2. If there is a number with an exponent, then evaluate the number with the exponent.

3. Multiplication and Division is performed before Addition and Subtraction.

4. If there is Multiplication or Division next to each other, do whichever is written first (left to right).

5. If there is Addition or Subtraction next to each other, do whichever is written first (left to right).

Before we perform order of operations with fractions, let's practice a few problems of order of operations with whole numbers. The Order of Operations is a priority list. The operations higher up on the priority list get taken care of first.

Example 1a. Simplify $13 + 2 \cdot 3$ Step 1. Multiply

$\qquad\qquad\qquad 13 + 6$ Step 2. Add

Answer: $\boxed{19}$

Example 1b: Simplify $4 + 3(12 - 7)$ Step 1. Parentheses

$4 + 3(5)$ Step 2. Multiply

$4 + 15$ Step 3. Add

Answer: $\boxed{19}$

Your Turn Problem #1

a) Simplify $8 + 12 \div 3$ b) Simplify $5 + 3(7 - 2)$

Answer:_____   Answer:_____

Example 2a. Simplify $4^2 \div (7 - 5)$ Step 1. Parentheses

$4^2 \div (2)$ Step 2. Exponents

$16 \div (2)$ Step 3. Division

Answer: $\boxed{8}$

Example 2b. Simplify $(5 - 2)^2 + (8 - 6)^3$ Step 1. Parentheses

$3^2 + 2^3$ Step 2. Exponents

$9 + 8$ Step 3. Add

Answer: $\boxed{17}$

Your Turn Problem #2

a) Simplify $(15 - 9)^2 \div 4$ b) Simplify $(11 - 3)(5 + 2)$

Answer:_____   Answer:_____

Example 3. Simplify $1\frac{2}{5} \div \left(\frac{1}{4} + \frac{3}{5} \right)$

Solution: First, do inside the parentheses. Add fractions. Then recopy.

$$\frac{1}{4} \cdot \frac{5}{5} = \frac{5}{20}$$
$$+ \quad \frac{3}{5} \cdot \frac{4}{4} = \frac{12}{20}$$
$$\overline{\phantom{+ \quad \frac{3}{5} \cdot \frac{4}{4} =} \quad \frac{17}{20}}$$

$$1\frac{2}{5} \div \frac{17}{20}$$

Division with mixed numbers: Change mixed numbers to improper fractions.

Then, convert to multiplication, reduce and multiply.

$$\frac{7}{5} \div \frac{17}{20} \Rightarrow \frac{7}{5} \cdot \frac{20}{17}$$

$$\Rightarrow \frac{7}{\cancel{5}} \cdot \frac{2 \cdot 2 \cdot \cancel{5}}{17} = \frac{28}{17}$$

Answer: $\boxed{1\frac{11}{17}}$

Your Turn Problem #3

Simplify: $\frac{1}{2} \div \frac{9}{8} + \frac{1}{2} \cdot \frac{2}{3}$

Example 4. Simplify $\left(3\dfrac{4}{15} - 2\dfrac{4}{5}\right) \div 3$

Solution: First, do inside the parentheses. Subtract mixed numbers.

$$3\dfrac{4}{15} \cdot \dfrac{1}{1} = \overset{2}{\cancel{3}}\dfrac{4}{15} + \dfrac{15}{15} = 2\dfrac{19}{15}$$
$$-\quad 2\dfrac{4}{5} \cdot \dfrac{3}{3} = 2\dfrac{12}{15} \quad = 2\dfrac{12}{15}$$
$$\rule{4cm}{0.4pt}$$
$$\dfrac{7}{15}$$

Recopy problem and replace using the $\dfrac{7}{15}$.

$$\dfrac{7}{15} \div 3 \;\Rightarrow\; \dfrac{7}{15} \div \dfrac{3}{1}$$

$$\Rightarrow\; \dfrac{7}{15} \cdot \dfrac{1}{3}$$

Answer: $\boxed{\dfrac{7}{45}}$

Your Turn Problem #4

Simplify: $1\dfrac{2}{3} \div \left(5 - 4\dfrac{1}{3}\right)$

Example 5. Simplify $\left(2\dfrac{7}{12}-1\dfrac{1}{3}\right)\div\left(2\dfrac{1}{2}\right)^2$

Solution: 1. Do inside the parentheses. Subtract the mixed numbers.

$$2\dfrac{7}{12}\cdot\dfrac{1}{1}=2\dfrac{7}{12}$$

$$-\ 1\dfrac{1}{3}\cdot\dfrac{4}{4}=\ 1\dfrac{4}{12}$$

$$1\dfrac{3}{12}=1\dfrac{1}{4}$$

2. Replace and rewrite. Then do exponents.

$$\left(1\dfrac{1}{4}\right)\div\left(2\dfrac{1}{2}\right)^2$$

$$\left(2\dfrac{1}{2}\right)^2=2\dfrac{1}{2}\cdot 2\dfrac{1}{2}=\dfrac{5}{2}\cdot\dfrac{5}{2}=\dfrac{25}{4}$$

3. Replace and rewrite. Then do the division.

$$1\dfrac{1}{4}\div\dfrac{25}{4}$$

$$\Rightarrow\dfrac{5}{4}\div\dfrac{25}{4}$$

$$\Rightarrow\dfrac{5}{4}\cdot\dfrac{4}{25}$$

Answer: $\boxed{\dfrac{1}{5}}$

Your Turn Problem #5

Simplify: $\left(8-7\dfrac{1}{3}\right)^2\div 4$

6.6 Homework: Order of Operations with Fractions

Simplify using the order of operations.

1. $\dfrac{1}{12} + \dfrac{3}{4} \cdot \dfrac{5}{9}$

2. $8 + 2\dfrac{1}{2} \cdot 1\dfrac{1}{2}$

3. $\dfrac{1}{2} + 5\dfrac{4}{5} \div 9\dfrac{2}{3}$

4. $\left(\dfrac{2}{3} + \dfrac{3}{4}\right) \div \left(\dfrac{5}{6} - \dfrac{1}{3}\right)$

5. $15\dfrac{1}{9} - \left(8\dfrac{2}{3} \div 3\dfrac{3}{4}\right)$

6. $10 - 5\dfrac{2}{3} \div 1\dfrac{2}{3} + 8\dfrac{1}{2}$

6.6 Homework: Order of Operations with Fractions

Simplify using the order of operations.

7. $\left(\dfrac{3}{4}\right)^3$

8. $\left(3\dfrac{2}{5}\right)^2$

9. $\left(\dfrac{2}{3}\right)^2 + \left(\dfrac{1}{2}\right)^3$

10. $\left(2\dfrac{4}{9} + 1\dfrac{1}{18}\right) \div \left(1\dfrac{2}{9} - \dfrac{1}{6}\right)$

11. $5\dfrac{1}{7} \div \left(2 + 1\dfrac{1}{3}\right)^2$

12. $\left(19\dfrac{1}{3} - 18\dfrac{3}{4}\right) \div \left(\dfrac{1}{3} + \dfrac{1}{4}\right)$

Practice Test 6

Find the LCM for the following

1. 15, 55

2. 12, 20, 35

Perform the indicated operation. Do not leave improper fractions for answers.

3. $\dfrac{5}{16} + \dfrac{3}{16}$

4. $\dfrac{2}{9} + \dfrac{4}{15}$

5. $\dfrac{13}{20} + \dfrac{7}{12}$

6. $\dfrac{7}{24} + \dfrac{9}{16} + \dfrac{5}{18}$

7. $\dfrac{7}{18} - \dfrac{5}{18}$

8. $\dfrac{7}{9} - \dfrac{2}{3}$

9. $\dfrac{11}{15} - \dfrac{5}{12}$

10. $\dfrac{13}{14} - \dfrac{8}{21}$

Practice Test 6 cont.

Perform the indicated operation. Do not leave improper fractions for answers.

11. $19\dfrac{5}{12}+15\dfrac{11}{12}$

12. $115\dfrac{3}{8}+43\dfrac{2}{5}$

13. $26\dfrac{3}{4}+18\dfrac{11}{20}$

14. $17\dfrac{2}{5}+\dfrac{3}{4}$

15. $17\dfrac{2}{3}+5\dfrac{1}{2}+3\dfrac{3}{4}$

16. $17\dfrac{5}{12}+11\dfrac{2}{3}+\dfrac{7}{8}$

17. $85\dfrac{4}{9}-25\dfrac{1}{9}$

18. $17\dfrac{3}{10}-9\dfrac{5}{6}$

19. $43\dfrac{5}{16}-27\dfrac{9}{10}$

20. $15-2\dfrac{5}{13}$

Practice Test 6 cont.

Solve each application. Do not leave improper fractions for answers.

21. Find the perimeter of a rectangle if each width is $\frac{1}{3}$ yd and each length is $\frac{7}{12}$ yd.

22. Find the perimeter of the triangle if the lengths of the sides are $11\frac{7}{12}$ ft, $5\frac{2}{3}$ ft, and $8\frac{3}{4}$ ft.

23. Nicholas lives $\frac{7}{8}$ of a mile from school. After he walks $\frac{1}{2}$ mile, how much further is it to school?

24. A person spent $\frac{2}{5}$ of a day sleeping and $\frac{1}{3}$ of a day working. What is the total fraction of the day spent sleeping or working?

25. In Running Springs, they received $11\frac{1}{4}$ inches of snow in December, $15\frac{2}{3}$ inches of snow in January, and $14\frac{1}{2}$ inches of snow in February. What is the total amount of snowfall for the three months?

Practice Test 6 cont.

Solve each application. Do not leave improper fractions for answers.

26. A roast weighed $4\frac{3}{8}$ lb before cooking and $3\frac{11}{16}$ lb after cooking. How many pounds were lost during the cooking?

27. A roll of cable contains 50 yards. If $18\frac{2}{3}$ yards is cut from the roll, how much cable remains?

28. A technician needs pieces of fiber optic cable $24\frac{5}{8}$ ft and $32\frac{2}{3}$ ft long. What is the total length of cable that is needed?

29. The average person on salary spends $6\frac{4}{15}$ hours sleeping and $9\frac{1}{4}$ hours at work. What is the total number of the hours per day spent sleeping or at work?

7.1 Introduction to Decimals

Decimal Places

Digits located to the right of the decimal point occupy *decimal places.* The decimal places are represented by powers of 10 that would appear in the denominator of equivalent fractions written in fraction form.

The first place to the right of the decimal point is the *tenths place* because it is representative of the common fraction 1/10. Likewise, the second place to the right of the decimal point is the *hundredths place* because it is representative of the common fraction 1/100.

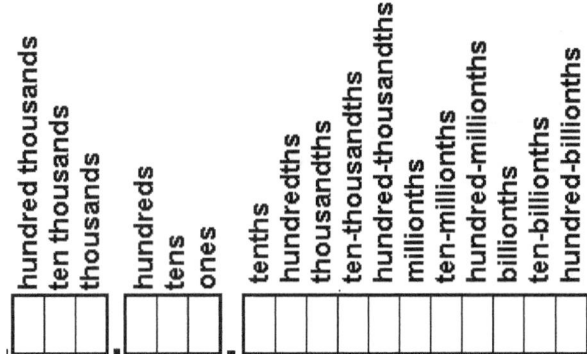

Consider the monetary value: $8.73.

This means 8 one dollar bills (ones), 7 dimes (seven-tenths of a dollar), and 3 pennies (3 hundredths of a dollar.)

73.652:

The decimal notation 73.652 means 7 tens+3 ones+6 tenths+5 hundredths+2 thousandths.

Writing a decimal in words

Procedure: Writing a Decimal in Word Form

Step 1.   Write the whole number part in words (the number to the left of the decimal point).

Step 2.   Write the word "and" in place of the decimal point.

Step 3.   Write the decimal part in words as though it were a whole number without any commas *followed by the name of the place value of the last digit.*

Example 1. Write the following in words.   **Answers:**

a) 0.083   **a)** eighty-three thousandths

b) 7.0304   **b)** seven and three hundred four ten-thousandths

c) 12.52   **c)** twelve and fifty-two hundredths

Your Turn Problem #1

Write the following in words.

a) 0.75

b) 91.036

Changing a decimal written in words to decimal notation.

When writing decimals in decimal notation, make sure the last digit is in the correct place value.

Example 2: Write the following in decimal notation

a) Five hundredths

The 5 must be in the hundredths place.

Answer: 0.05

b) forty-seven and twelve ten-thousandths

The words before "and" give the whole number in front of the decimal point. 47 will be written before the decimal point. 12 will be written after the decimal point. Place enough zeros before the 12 so that the last digit of the 12 is in the ten-thousandths place.

Answer: 47.0012

Your Turn Problem #2

Write the following in words: 58.943

Writing a Fraction as a decimal where the denominator is a power of 10

If the denominator is power of 10, such as 10, 100, 1000, etc., then its decimal equivalent will have the same number of places as there are zeros. The last digit will be in the place value of the same number in the denominator. If there is a whole number in written in front of the fraction, then the same whole number will be written in front of the decimal point. If there is no whole number in front of the fraction, then a zero should be written in front of the decimal point.

Denominator = 10, $\Rightarrow$ last digit in tenths place.

Denominator = 100, $\Rightarrow$ last digit in hundredths place.

Denominator = 1000, $\Rightarrow$ last digit in thousandths place.

etc.

Example 3. Write as a decimal

a) $\dfrac{3}{10}$ b) $\dfrac{4}{100}$ c) $\dfrac{27}{1000}$ d) $19\dfrac{23}{100}$

Answers: a) $\dfrac{3}{10} = 0.3$ b) $\dfrac{4}{100} = 0.04$ c) $\dfrac{27}{1000} = 0.027$ d) $19\dfrac{23}{100} = 19.23$

Your Turn Problem #3

Write as a decimal

a) $\dfrac{7}{10}$ b) $\dfrac{13}{100}$ c) $67\dfrac{15}{1000}$

Converting a Decimal to a Fraction or Mixed Number

Procedure: To convert a decimal to a fraction or mixed number:

1. If the decimal has a whole number in front of the decimal point, then this whole number will also be the whole number part of the mixed number.

2. The digits to the right of the decimal point will be the numerator of the fraction.

3. The denominator will be the place value of the last digit. If the last digit of the decimal is in the tenths place, the denominator will be ten. If the last digit of the decimal is in the hundredths place, the denominator will be one hundred. If the last digit of the decimal is in the thousandths place, the denominator will be one thousand. And so on...

Example 4. Write the following as fractions. (Fractions should always be reduced if possible.)

a) 0.8   b) 0.25   c) 81.037

Answers: a) $0.8 = \dfrac{8}{10} = \dfrac{4}{5}$ b) $0.25 = \dfrac{25}{100} = \dfrac{1}{4}$ c) $81.037 = 81\dfrac{37}{1000}$

Your Turn Problem #4

Write the following as fractions. Reduce if possible.

a) 0.4   **b)** 0.52   **c)** 12.06

Rounding Decimals

Procedure: To round to a certain place:

a) Locate the digit in the place value which is to be rounded.

b) Consider the next digit to the right.

c) If the digit to the right is 5 or greater, round up; add 1 to the specified place value.
   If the digit to the right is 4 or lower, round down; the specified place value remains the same.

When rounding to a specific place, the last digit written should be the specified place value. Do not write any digits past that value.

Example 5. Round the following to the tenths place.

      a) 0.274   b) 5.452   c) 16.395   d) 7.982

Answers: a) 0.3   b) 5.5   c) 16.4   d) 8.0

Your Turn Problem #5

Round the following to the tenths place.

a) 9.583   b) 205.797   c) 0.81349

Example 6. Round the following to the hundredths place.

      a) 0.5749   b) 3.2864   c) 54.397   d) 9.994

Answers: a) 0.57   b) 3.29   c) 54.40   d) 9.99

Your Turn Problem #6

Round the following to the hundredths place.

a) 7.2548   b) 71.997   c) 0.7829

Example 7. Round the following to the indicated place value.

a) 8675.3087 (thousandths)   b) 0.0052384 (millionths)   c) 8.9244 (ones)

Answers: a) 8675.309   b) 0.005238   c) 9

Your Turn Problem #7

Round the following to the indicated place value.

a) 5.2858 (thousandths)   b) 0.071287 (ten-thousandths)   c) 19.97 (ones)

7.1 Homework: Introduction to Decimals

Write in Words.

1. 0.8

2. 0.25

3. 0.044

4. 12.3

5. 15.87

6. 43.034

Write in decimal form.

7. Eight tenths

8. Twenty-four hundredths

9. Thirty-six thousandths

10. Twelve and four tenths

11. Nine and fifty-four thousandths

12. Seven thousandths

13. Five and one hundred twenty-two thousandths

14. Forty-three thousandths

15. One and seven hundred thirty-two thousandths

16. Two and two thousandths

17. Five hundred sixty-seven thousandths

7.1 Homework Cont.

Write each of the following as a proper fraction or a mixed number. Reduce if possible.

18. 0.4

19. 0.25

20. 0.091

21. 3.8

22. 5.21

23. 16.003

24. 27.125

Write in decimal notation

25. $\dfrac{3}{10}$

26. $\dfrac{29}{100}$

27. $\dfrac{37}{1000}$

28. $12\dfrac{8}{100}$

29. $34\dfrac{17}{1000}$

30. $79\dfrac{3}{10}$

7.1 Homework Cont.

Round to the nearest tenth.

31. 0.46

32. 0.253

33. 0.091

34. 3.882

35. 5.971

36. 16.0675

Round to the nearest hundredth.

37. 0.468

38. 0.253

39. 0.915

40. 3.8826

41. 5.9949

42. 16.0675

7.1 Homework Cont.

Round to the nearest thousandth.

43. 0.4682

44. 0.2539

45. 0.9155

46. 3.8826

47. 5.9949

48. 16.0675

Round to the indicated place value:

49. 0.596 ; hundredths

50. 2.92986 ; ten-thousandths

51. 15.9522 ; tenths

52. 7.28676 ; thousandths

53. 0.0426 ; hundredths

54. 9.532 ; ones

55. 314.9072 ; hundredths

56. 79.612049 ; ten-thousandths

57. 0.498 ; ones

7.2 Adding and Subtracting with Decimals

Adding and Subtracting with decimals is similar to adding and subtracting with whole numbers.

Procedure: Adding and Subtracting with Decimals

1. Write the problem vertically aligning the decimal points. Add zeros if necessary.

2. Perform the operation. (Add or Subtract)

3. Place the decimal point in the sum or difference so that it is vertically aligned with the decimal points above.

Example 1. Add: $17.672 + 5.65$

$$
\begin{array}{r}
1\,7\,.\,6\,7\,2 \\
+5\,.\,6\,5\,0 \\
\hline
2\,3\,.\,3\,2\,2
\end{array}
$$

Step 1. Write vertically, aligning the decimal points.
Step 2 &3. Add, and place the decimal point so it is vertically aligned.

Your Turn Problem #1

Add: $12.2 + 5.73$

Example 2. Subtract: $45.3 - 7.94$

$$
\begin{array}{r}
4\,5\,.\,3\,0 \\
-7\,.\,9\,4 \\
\hline
3\,7\,.\,3\,6
\end{array}
$$

Step 1. Write vertically, aligning the decimal points.
Step 2 &3. Subtract, and place the decimal point so it is vertically aligned.

Your Turn Problem #2

Subtract: $8 - 3.75$
(Hint: $8 = 8.00$)

Example 3. Find the perimeter of a rectangle with a length of 7.5 inches and a width of 3.4 inches.

Solution: A rectangle has 4 sides. We need to add all 4 sides. The order does not matter since addition is commutative.

$$7.5 \,in + 7.5 \,in + 3.4 \,in + 3.4 \,in$$

$$
\begin{array}{r}
7.5 \\
+\,7.5 \\
\hline
15.0
\end{array}
\qquad
\begin{array}{r}
3.4 \\
+\,3.4 \\
\hline
6.8
\end{array}
\qquad
\begin{array}{r}
15.0 \\
+\,6.8 \\
\hline
21.8
\end{array}
$$

Answer: The perimeter of the rectangle is 21.8 in.

Your Turn Problem #3

Raul bought a Digital Camera for $84.48 with tax. If he paid with $100 (a hundred dollar bill), how much change will he receive?

7.2 Homework: Adding and Subtracting with Decimals

Perform the indicated operation.

1. 14.62+5.3

2. 18+4.25

3. 7.21+0.35

4. 0.54+2.85

5. $542.36+$79.21

6. 9.8+5.78+0.321

7. 5+1.03+0.29

8. 21+0.45

9. 0.98+0.75

10. 9.75−5.32

11. 18.8−6.52

12. 8.00−5.43

7.2 Homework Cont.

Perform the indicated operation.

13. $15-2.79$

14. $5-1.154$

15. $\$25-\$15.32$

16. Find the perimeter of a rectangle with a length of 22.5 inches and a width of 12.2 inches.

17. Erin's monthly pay check was for $2,362.36. The mortgage was $725.26, the utilities added up to $276.78, and her insurance payment was $64.26. How much does she have left?

18. On a 4-day trip, Aimee bought 12.2, 11.5, 7.9, and 12.8 gallons of gas. How many gallons of gas did she buy?

7.2 Homework Cont.

19. Erika bought a Sony T.V. selling for $475.50 marked down to $299.75. What was her savings?

20. Sandra walked 6.31 miles on Monday, 8.56 miles on Wednesday, and 3.26 miles on Friday. How many miles did she walk for the three days?

21. In the 1996 Olympics, the Cuban mile relay team ran laps of 48.76 seconds, 49.38 seconds, 47.90 seconds, and 46.16 seconds. What was their total time?

22. Fernando's rent on his apartment was raised from $818.63 per month to $903.96 per month. How much more per month must he spend on rent?

23. Dominic bought a new edger for $74.99. If he gave the clerk a $100 bill, how much change will he receive?

7.3 Multiplying with Decimals

Multiplying numbers where at least one the numbers is a decimal.

Procedure: To multiply numbers in which at least one of the numbers is a decimal:
Step 1. Align the numbers vertically and multiply as if whole numbers.
Step 2. Find the sum of the decimal places by counting the digits to the right of the decimal point in each number, and then adding together.
Step 3. Starting to the right of the answer obtained from multiplying, (Step 1), count from right to left the same number as the sum of the decimal places determined in Step 2. Insert the decimal point.

Example 1. Multiply: 0.111×0.7

$$
\begin{array}{r}
0.111 \quad \leftarrow 3 \text{ places} \\
\times \quad 0.7 \quad \leftarrow 1 \text{ place} \\
\hline
777 \quad \leftarrow 4 \text{ places}
\end{array}
$$

Move the decimal point 4 places to the left starting after the last digit.

Answer: $\boxed{0.0777}$

Your Turn Problem #1
Multiply 0.264×0.23

Example 2. Find the product: 4.36×14.8

Solution:

$$
\begin{array}{r}
4.36 \quad \leftarrow 2 \text{ places} \\
\times 14.8 \quad \leftarrow \underline{1 \text{ place}} \\
\hline
3488 \qquad 3 \text{ places} \\
17440 \\
\underline{43600} \\
64528
\end{array}
$$

Start the decimal point after the 8. Move it 3 places to the left.

Answer: $\boxed{64.528}$

Your Turn Problem #2

Multiply 5.95×0.078

Multiplying by Powers of 10

Procedure for Multiplying by Powers of 10, (i.e. 10, 100, 1000 etc.)

Move the decimal point to the right the same number of places as there are zeros in the power of 10.

Power of 10: A power of 10 is any number that results from raising the number "10" to any whole number exponent.

Examples are: $10^1 = 10$

$10^2 = 100$

$10^3 = 1,000$ (The number of zeros is the same as the exponent.)

$10^4 = 10,000$

$10^5 = 100,000$

Example 3a. Multiply 6.37×1000

Solution: Since there are 3 zeros, move the decimal point to the right 3 places. To move the decimal 3 places to the right, add a zero to have 3 places.

Answer: $\boxed{6370}$

Example 3b. Multiply 7.295×10^2

Solution: $10^2 = 10 \times 10 = 100$. So there are 2 zeros (same as exponent). Move the decimal point to the right 2 places.

Answer: $\boxed{729.5}$

Your Turn Problem #3

Multiply the following: a) 0.843×100 b) 0.8675309×10^3

338

7.3 Homework: Multiplying with Decimals

Perform the following operations.

1. 11.2×4.6

2. 5.9×3.3

3. 1.56×91

4. 12.5×4.3

5. 15×0.812

6. 0.436×4.2

7. 0.017×0.02

8. 2.375×0.022

9. 15.52×10

10. 1.402×100

11. 2.5×100

12. 0.12561×1000

13. 5.54×1000

14. 0.432×10^2

15. 48.131×10^3

7.3 Homework Cont.

Solve the following application problems.

16. Jose makes monthly payments of $214.56 on his new car. What will his payments be for one year?

17. Mesquite turkey meat at a deli cost $4.75 per pound. How much would 3.5 lbs cost? (Round to the nearest cent if necessary.)

18. If Anna earns $12.60 per hour, what would her overtime pay be if overtime = time and a half? (Multiply. by 1.5)

19. Anna earns $12.60 per hour. For overtime (each hour over 40), she earns time and a half. If she worked 48 hours, what pay should she receive?

Practice Test 7

Write in Words.

1. 0.035

2. 82.41

Write in decimal form.

3. Nineteen and two hundred fifty-three thousandths

4. One and thirty-two thousandths

Write each of the following as a proper fraction or a mixed number. Reduce if possible.

5. 9.4

6. 5.35

Write in decimal notation

7. $\dfrac{17}{100}$

8. $44\dfrac{19}{1000}$

Round to the indicated place value:

9. 5.9385 ; tenths

10. 16.9488 ; tenths

11. 0.4283 ; hundredths

12. 12.4972 ; hundredths

13. 867.5309 ; thousandths

14. 3.14159 ; ten-thousandths

Practice Test 7 cont.

Perform the indicated operation.

15. $8.46 + 9.8$

16. $\$299.65 + \$89.99$

17. $7 + 2.478 + 0.365$

18. $25.4 - 8.78$

19. $20 - 4.36$

20. $\$40 - \$24.86$

21. 78.6×6.8

22. 0.455×0.24

23. 0.0314×100

24. 5.23×1000

25. Find the perimeter of a rectangle with a length of 14.6 meters and a width of 8.7 meters.

Practice Test 7 cont.

26. Katie's monthly pay check was for $3,422.65. The mortgage was $1,275.26, the utilities added up to $197.88, and her insurance payment was $73.76. How much does she have left?

27. In the 2009 World Championship, the US team consisting of Piersol, Shanteau, Phelps, and Walters swam the 4 X 100 meter medley with times of 52.19 seconds, 58.57 seconds, 49.72 seconds, and 46.80 seconds respectively. What was their total time?

28. Alberto's mortgage payment on his house decreased from $2,587.98 per month to $1,903.54 per month. How much less per month is he paying on his mortgage?

29. Jimmy bought a new backpack for $45.27. If he gave the clerk a $100 bill, how much change will he receive?

Practice Test 7 cont.

30. Kendra makes monthly payments of $457.56 on her new car. What will her payments be for one year?

31. Honey Crisp apples at a market cost $1.49 per pound. How much would 2.3 lbs cost? Round to the nearest cent if necessary.

32. If Yolanda earns $14.80 per hour, what would her overtime wage be if overtime = time and a half? (Multiply by 1.5)

33. Yolanda earns $14.80 per hour. For overtime (each hour over 40), she earns time and a half. If she worked 43 hours, what pay should she receive?

8.1 Dividing with Decimals

Terminology

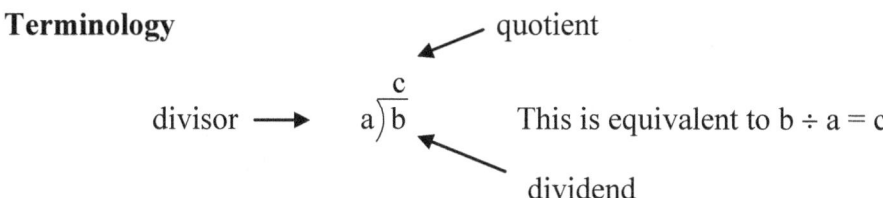

quotient

divisor ⟶ $a\overline{)b}^{\,c}$ This is equivalent to $b \div a = c$

dividend

Dividing by a Whole Number

To divide by a whole number (i.e., the divisor is a whole number)

Step 1. Rewrite the problem in long division form.

Step 2. Place the decimal point in the quotient directly above the decimal point in the dividend and divide.

Step 3. If necessary, write additional zeros to the right of the last digit following the decimal point in the dividend to allow the division to continue.

Example 1. Divide: $29.4 \div 12$

$$
12\overline{)29.4} \;\rightarrow\;
\begin{array}{r}
2.45 \\
12\overline{)29.40} \\
-24 \\
\hline
54 \\
-48 \\
\hline
60 \\
-60 \\
\hline
0
\end{array}
$$

Step 1. Rewrite in long division form.

Step 2. Write the decimal point straight up. Then divide. The first number in the quotient will be written above the last digit of the number you are dividing into. In this example, the 2 is written directly above the 9 since we start by dividing 12 into 29.

Answer: $\boxed{2.45}$

Your Turn Problem #1

Divide: $59.55 \div 15$

Dividing by a Decimal

To divide by a decimal (i.e., the divisor is not a whole number)

Step 1. Rewrite the problem in long division form.

Step 2. Move the decimal to the right (counting the places) to make the divisor a whole number.

Then move the decimal point in the dividend the same number of places to the right.

Step 3. Rewrite the problem and divide.

Example 2. Divide: $0.5072 \div 0.08$

$$0.08\overline{)0.5072} \rightarrow 8\overline{)50.72}$$

$$
\begin{array}{r}
6.34 \\
8\overline{)50.72} \\
-48 \\
\hline
27 \\
-24 \\
\hline
32 \\
-32 \\
\hline
0
\end{array}
$$

Step 1. Rewrite in long division form.

Step 2. Move the decimal in the divisor two places to the right to make it a whole number. Then move the decimal point in the dividend two places to the right.

Step 3. Write the decimal point straight up. Then divide. The first number in the quotient will be written above the last digit of the number you are dividing into. In this example, the 6 is written directly above the 0 since we begin by dividing 8 into 50.

Answer: $\boxed{6.34}$

Your Turn Problem #2

Divide: $3.5973 \div 0.07$

Example 3. Divide: $10.86 \div 1.5$

$1.5 \overline{)10.86} \rightarrow 15 \overline{)108.6}$

$$
\begin{array}{r}
7.24 \\
15 \overline{)108.60} \\
-105 \\
\overline{36} \\
-30 \\
\overline{60} \\
-60 \\
\overline{0}
\end{array}
$$

Step 1.   Rewrite in long division form.

Step 2.   Move the decimal in the divisor one place to the right to make it a whole number. Then move the decimal point in the dividend one place to the right.

Step 3.   Write the decimal point straight up. Then divide. We need to add a zero after the 6 to keep dividing until it divides evenly without a remainder.

Answer: 7.24

Your Turn Problem #3

Divide: $12.25 \div 1.4$

The answers in the previous examples were all exact values. In other words, the division ended. This will not always occur. Sometimes, the quotient will continue without end. There will always be a remainder. In that case, we will round to a specified place value and disregard the remainder.

<u>Rounding the quotient to a specified place value.</u>

To round a decimal, we look at the number to the right of the place value of which we are rounding.

If 5 or more, round up. Add 1 to the specified place value.

If 4 or less, round down. Leave the specified place value alone.

In both cases, the last digit written is the specified place value being rounded to.

Rule: When rounding a quotient to a specified place value:

We need to obtain one more place value in the quotient to the right of the specified place value to round.

Examples: If rounding to the tenths place, you must go to the hundredths. If rounding to the hundredths place, you must go to the thousandths. Etc.

Example 4. Divide: $7.2 \div 7$ (round to the hundredths place)

$$7\overline{)7.200} \rightarrow$$

$$
\begin{array}{r}
1.028 \\
7\overline{)7.200} \\
-7 \\
\hline
020 \\
-14 \\
\hline
60 \\
-56 \\
\hline
4
\end{array}
$$

Step 1. Rewrite in long division form.

Step 2. Write the decimal point straight up. Then divide. We need to add two zeros after the 2 to get to the thousandths place.

Step 3. Then round answer to the hundredths place.

Answer: $\boxed{1.03}$

Your Turn Problem #4

Divide: $8.3 \div 1.1$ (round to the hundredths place)

Example 5. Divide: $0.07 \div 1.3$ (round to the thousandths place)

$$1.3\overline{)0.07} \rightarrow 13\overline{)0.7000}$$

Step 1. Rewrite in long division form.

Step 2. Move the decimal in the divisor one place to the right to make it a whole number. Then move the decimal point in the dividend one place to the right.

$$\begin{array}{r} 0.0538 \\ 13\overline{)0.7000} \\ -65 \\ \hline 50 \\ -39 \\ \hline 110 \\ -104 \\ \hline 6 \end{array}$$

Step 3. Write the decimal point straight up. Then divide. We need to add three zeros after the 7 to get to the ten-thousandths place.

Step 4. Then round answer to the thousandths place.

Answer: $\boxed{0.054}$

Your Turn Problem #5

Divide: $0.08 \div 1.3$ (round to the thousandths place)

Dividing by Powers of 10

Procedure for Dividing by Powers of 10, (i.e. 10, 100, 1000 etc.)

When dividing by powers of 10, move the decimal point to the left the same number of places as there are zeros in the power of 10. This is the quotient (answer).

Examples: 10 has one zero. If a number is being divided by 10, move the decimal point one place to the left.

100 has two zeros. If a number is being divided by 100, move the decimal point two places to the left.

Example 6a. Divide $6.37 \div 1000$

Solution: Since there are 3 zeros, move the decimal point to the left 3 places. To move the decimal 3 places to the left, add two zeros to have 3 places.

Answer: $\boxed{0.00637}$

Example 6b. Divide $729 \div 10^2$

Solution: $10^2 = 10 \times 10 = 100$. So there are 2 zeros (same as exponent). Move the decimal point to the left 2 places.

Answer: $\boxed{7.29}$

Your Turn Problem #6

Divide the following: a) $5.2 \div 100$ b) $8675309 \div 10^4$

8.1 Homework: Dividing with Decimals

Perform the following operations.

1. $14.22 \div 6$

2. $1.88 \div 8$

3. $88.494 \div 14$

4. $65.052 \div 52$

5. $10.92 \div 1.2$

6. $0.3556 \div 1.4$

7. $4.2714 \div 2.1$

8. $0.67782 \div 0.13$

8.1 Homework cont.

Perform the following operations.

9. $0.04984 \div 0.008$

10. $0.00552 \div 0.24$

11. $1.2972 \div 0.023$

12. $3.10432 \div 0.16$

13. $0.2568 \div 0.012$

14. $7.14 \div 0.3$

15. $8.17 \div 0.08$

16. $1.23 \div 0.24$

8.1 Homework cont.

Perform the following operations.

17. $15.52 \div 10$ **18.** $140.2 \div 100$

19. $2.5 \div 100$ **20.** $12561 \div 1000$

21. $5.54 \div 100$ **22.** $432.6 \div 10^2$

23. $48.131 \div 10^3$ **24.** $57.8 \div 10^4$

Perform the following operations. Round to the indicated place value.

25. $3.46 \div 0.7$ hundredth   **26.** $45 \div 1.3$ hundredth

8.1 Homework cont.

Perform the following operations. Round to the indicated place value.

27. $8 \div 1.3$ tenth

28. $5.4 \div 21$ hundredth

29. $54 \div 0.7$ thousandth

30. $5.94 \div 0.7$ hundredth

Solve the following application problems.

31. Rodney paid \$37.23 for three CD's on sale. If each CD were equal in price, what was the cost per CD?

32. Manuel purchased a Kenmore refrigerator for \$742.08. If he pays it off in 12 equal payments, what amount will he be paying per month?

33. The pollution index readings for a three day period were 23.4, 21.5, and 17.6. What was the average reading (to the nearest 10th) for the three days?

8.2 Circumference and Area of a Circle

A circle is a plane figure that consists of all points that lie the same distance from a fixed point. The fixed point that defines a circle is called the center.

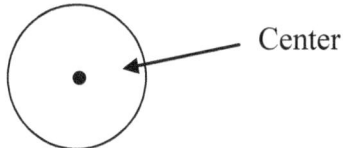 Center

Radius

A radius is a segment with one endpoint on the center and the other endpoint on the circle.

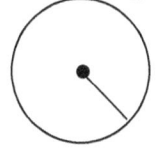 Radius

Diameter

A diameter is a segment that passes through the center of the circle and has endpoints on the circle.

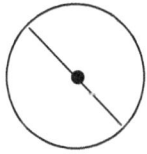
Diameter
The diameter is twice the radius.
The radius is ½ the diameter

Example 1. Given the diameter, find the radius

a) diameter = 12 ft   b) diameter = 28 in   c) diameter = 13 ft

Solution: To find the radius given the diameter, divide the diameter by 2.

Answers: a) radius = 6 ft   b) radius = 14 in   c) radius = 6.5 ft $(13 \div 2 = 6.5)$

Your Turn Problem #1

If the diameter is 21 feet, find the radius.

Example 2. Given the radius, find the diameter

    a) radius = 9 ft   b) radius = 23 in   c) radius = 7.5 ft

Solution: To find the diameter given the radius, multiply the radius by 2.

    a) diameter = 18 ft   b) diameter = 46 in   c) diameter = 15 ft $(7.5 \times 2 = 15)$

Your Turn Problem #2

If the radius is 18 feet, find the diameter.

Often in mathematics, we use formulas to find information. For example, we use a formula to find the area of a rectangle. The formula is $A = L \times W$, where L is the length of the rectangle and W is the width of the rectangle.

Example: Find the area of a rectangle if the length is 8 ft and the width is 5 ft.

          Using the formula, $A = L \times W$

$$A = 8\,\text{ft} \times 5\,\text{ft}$$

$$A = 40\,\text{ft}^2$$

Answer: The area of the rectangle is 40 square feet.

Area of a Circle

The formula will involve the Greek letter π. It is called pi (pronounced just like pie). The value of π is approximately equal to 3.14159.... Actually, the decimal representation of π never ends or repeats. For our purposes, we will use 3.14 for π.

To find the area of a circle, we will use the following formula: $A = \pi \cdot r^2$, where r is the radius. If r is the radius, r^2 means $r \times r$.

Example 3. Find the area of a circle if the radius is 8 ft.

Solution: Using the formula, $A = \pi \times r^2$, substitute the values of π and r. Then simplify using the order of operations. The units for area are usually square units, i.e., ft^2, m^2, in^2.

$$A = 3.14 \times (8\,\text{ft})^2$$
$$A = 3.14 \times 8\text{ft} \times 8\text{ft}$$
$$A = 3.14 \times 64\,\text{ft}^2$$
$$A = 200.96\,\text{ft}^2$$

Answer: The area of the circle is $200.96\,\text{ft}^2$.

Your Turn Problem #3

Find the area of a circle if the radius is 7 ft.

Example 4. Find the area of a circle if the diameter 15 ft.

Solution: Using the formula, $A = \pi \times r^2$, substitute the values of π and r.

If the diameter is 15 ft, then the radius is 7 ft $(15 \div 2 = 7.5)$

$$A = 3.14 \times (7.5 \text{ ft})^2$$

$$A = 3.14 \times (7.5 \text{ ft} \times 7.5 \text{ ft})$$

$$A = 3.14 \times 56.25 \text{ ft}^2$$

$$A = 176.625 \text{ ft}^2$$

Answer: The area of the circle is 176.625 ft^2.

Your Turn Problem #4

Find the area of a circle if the diameter is 26 ft.

Circumference of a circle

The circumference of a circle is the distance around the circle (similar to perimeter).

To find the circumference of a circle, we will use the formula $C = \pi \cdot d$ where $\pi = 3.14$ and d is the diameter.

Example 5. Find the circumference of a circle if the diameter is 17 ft.

Solution: Using the formula, $C = \pi \times d$, substitute the values of π and d.

$$C = 3.14 \times 17 \text{ ft}$$

$$C = 53.38 \text{ ft}$$

Answer: The circumference of the circle is 53.38 ft.

Your Turn Problem #5

Find the circumference of a circle if the diameter is 5 ft.

Example 6. Find the circumference of a circle if the radius is 11 ft.

Solution: We first need to find the diameter. Diameter = 22 ft $\left(11\,\text{ft} \times 2 = 22\,\text{ft}\right)$.

Using the formula, $C = \pi \times d$, substitute the values of π and d.

$$C = 3.14 \times 22\,\text{ft}$$
$$C = 69.08\,\text{ft}$$

Answer: The circumference of the circle is $69.08\,\text{ft}$.

Your Turn Problem #6

Find the circumference of a circle if the radius is 6.2 ft.

8.2 Homework: Circumference and Area of a Circle

Find the area of the circles with the given information. Use $A = \pi \times r^2$ **where** $\pi = 3.14$.

1. radius = 9 ft

2. radius = 12 ft

3. radius = 3.5 mi

4. diameter = 16 ft

5. diameter = 15 ft

6. diameter = 21 in

8.2 Homework Cont.

Find the circumference of the circles with the given information. Use C = π × d and π = 3.14.

7. diameter = 13 ft

8. diameter = 6 ft

9. diameter = 3.7 mi

10. radius = 13 ft

11. radius = 18 ft

12. radius = 4.3 in

8.3 Fractions and Decimals

Converting a Decimal to a Fraction or Mixed Number

Procedure: To convert a decimal to a fraction:

1. If the decimal has a whole number in front of the decimal point, this will also be the whole number part of the mixed number.

2. The digits to the right of the decimal point will be the numerator of the fraction.

3. The denominator will be the place value of the last digit.

Example 1. Write the following as fractions. (Fractions should always be reduced if possible.)

a) 0.8   b)   0.25   c) 81.037

Answers: a) $0.8 = \dfrac{8}{10} = \dfrac{4}{5}$ b) $0.25 = \dfrac{25}{100} = \dfrac{1}{4}$ c) $81.037 = 81\dfrac{37}{1000}$

Your Turn Problem #1

Write the following as fractions and reduce if possible.

a) 0.4   b) 0.52   c) 12.06

Writing a Fraction as a decimal where the denominator is a power of 10

If the denominator is power of 10, such as 10, 100, 1000, etc., then its decimal equivalent will have the same number of places as there are zeros. The last digit will be in the place value of the same number in the denominator. If there is a whole number in front of the fraction, then the same whole number will be the whole number in front of the decimal point.

Denominator = 10, $\Rightarrow$ last digit in tenths place.

Denominator = 100, $\Rightarrow$ last digit in hundredths place.

Denominator = 1000, $\Rightarrow$ last digit in thousandths place.

etc.

Example 2. Write as a decimal

     a) $\dfrac{3}{10}$ **b)** $\dfrac{4}{100}$ **c)** $\dfrac{27}{1000}$ **d)** $19\dfrac{23}{100}$

Answers: **a)** $\dfrac{3}{10} = 0.3$ **b)** $\dfrac{4}{100} = 0.04$ **c)** $\dfrac{27}{1000} = 0.027$ **d)** $19\dfrac{23}{100} = 19.23$

Your Turn Problem #2

Write the following fractions as decimals.

a) $\dfrac{7}{10}$ **b)** $\dfrac{13}{100}$ **c)** $67\dfrac{15}{1000}$

Converting fractions into a decimal where the decimal is non-terminating.

The result of the previous examples resulted in a *terminating decimal*. Any decimal whose ending digits do not continue indefinitely but that stop is a terminating decimal.

Examples of terminating decimals: 0.35, 0.625, 7.81

The decimal equivalents of some fractions are called repeating decimals. A repeating decimal is a representation of a fraction in which its ending digits repeat indefinitely with a pattern.

Examples of repeating decimals: 0.333333..., 0.434343...

We use a repeating decimal bar over the number or numbers that repeat to show that the pattern repeats over and over.

Examples: $0.3333.... = 0.\overline{3}$, $0.272727... = 0.\overline{27}$, $0.5888... = 0.5\overline{8}$

Note: In the last example, only the 8 is repeating. The bar must only be written above the 8. If you drew the bar above both the 5 and 8, it would be incorrect. The bar is only written over the number(s) that are being repeated.

Converting any Fraction to a Decimal

Procedure: To convert a fraction to a decimal:

Step 1: Rewrite as a long division problem with the denominator in front and the numerator inside the division box. $\left(\text{denominator} \,\overline{)\text{numerator}} \,\right)$

Step 2: Insert a decimal point after the last digit of the dividend (original numerator) and write at least one zero after the decimal point.

Step 3: Divide using the same steps as dividing decimals. Keep adding zeros to the dividend until you arrive at a zero remainder or until you start to see a repetition to a pattern of the digits in the quotient (the answer)

Note: If a repeating pattern emerges, then stop and write an over line above the last set of repeating digits.

Example 3. Convert $\dfrac{5}{8}$ to a decimal.

Solution:

1. Divide the denominator into the numerator.

2. Insert a decimal point after the 5. Bring the decimal point straight up in the quotient.

3. Divide. Keep going until it ends or there is a repeating pattern.

$$
\begin{array}{r}
0.625 \\
8\overline{)5.000} \\
-48 \\
\hline
20 \\
-16 \\
\hline
40 \\
-40 \\
\hline
0
\end{array}
$$

Answer: $\boxed{0.625}$

Your Turn Problem #3

Convert $\dfrac{3}{4}$ to a decimal.

Example 4. Convert $\dfrac{5}{18}$ to a decimal.

Solution:

1. Divide the denominator into the numerator.

2. Insert a decimal point after the 5. Bring the decimal point straight up in the quotient.

3. Divide. Keep going until it ends or there is a repeating pattern.

$$
\begin{array}{r}
0.277 \\
18\overline{)5.000} \\
-36 \\
\hline
140 \\
-126 \\
\hline
140 \\
-126 \\
\hline
14
\end{array}
$$

Answer: $\boxed{0.2\overline{7}}$

Note: As soon as we are dividing into the same number, the result will be a repeating decimal. Once 140 appeared for a 2nd time, the 7 was going to repeat.

Your Turn Problem #4

Convert $\dfrac{7}{11}$ to a decimal. Use a repeating decimal bar if necessary.

When converting a fraction to a decimal, the directions may state to round to an indicated place value. To round, we look at the number to the right of the indicated place value to determine whether to round up or down. So when rounding, we must always perform the division one place past that place value. If you are asked to round to the tenths place, divide until you have a number in the hundredths place. If you are asked to round to the hundredths place, divide until you have a number in the thousandths place. etc.

Example 5. Convert $\frac{3}{7}$ to a decimal. Round to the hundredths.

Solution:

1. Divide the denominator into the numerator.

2. Insert a decimal point after the 3. Bring the decimal point straight up in the quotient.

3. Divide. Add three zeros to get to the thousandths place. Then round to the hundredths.

$$
\begin{array}{r}
0.428 \\
7\overline{)3.000} \\
-28 \\
\hline
20 \\
-14 \\
\hline
60 \\
-56 \\
\hline
4
\end{array}
$$

Answer: $\boxed{0.43}$ (Round up since 8 is 5 or more.)

Your Turn Problem #5

Convert $\frac{7}{13}$ to a decimal. Round to the hundredths place.

Example 6. Convert $19\frac{1}{4}$ to a decimal.

Solution:

1. Divide the denominator into the numerator. The whole number, 19, will still be the whole number in front of the decimal point. So ignore the 19 until you are ready to write the final answer.

2. Insert a decimal point after the 1. Bring the decimal point straight up in the quotient.

3. Divide. Keep going until it ends or there is a repeating pattern.

$$
\begin{array}{r}
0.25 \\
4\overline{)1.000} \\
-8 \\
\hline
20 \\
-20 \\
\hline
0
\end{array}
$$

Answer: $\boxed{19.25}$

Your Turn Problem #6

Convert $31\dfrac{1}{8}$ to a decimal.

Example 7. Convert $75\dfrac{11}{18}$ to a decimal. Use a repeating decimal bar if necessary.

Solution:

1. Divide the denominator into the numerator. The whole number, 75, will still be the whole number in front of the decimal point. So ignore the 75 until you are ready to write the final answer.

2. Insert a decimal point after the 1. Bring the decimal point straight up in the quotient.

3. Divide. Keep going until it ends or there is a repeating pattern.

$$
\begin{array}{r}
0.611 \\
18\overline{)11.000} \\
-108 \\
\hline
20 \\
-18 \\
\hline
20 \\
-18 \\
\hline
2
\end{array}
$$

Answer: $\boxed{75.6\overline{1}}$

(The repeating decimal bar is only over the 1.)

Your Turn Problem #7

Convert $21\dfrac{5}{11}$ to a decimal. Use a repeating decimal bar if necessary.

8.3 Homework: Fractions and Decimals

Write each of the following as a proper fraction or a mixed number. Reduce if possible.

1. 0.4

2. 0.25

3. 0.091

4. 3.8

5. 5.21

6. 16.003

7. 27.125

8 71.6

9. 23.48

10. 101.05

11. 13.025

12. 8.15

8.3 Homework Cont.

Find the decimal equivalent for each of the following fractions.

13. $\dfrac{1}{4}$

14. $\dfrac{3}{4}$

15. $\dfrac{3}{5}$

16. $\dfrac{9}{10}$

17. $\dfrac{3}{8}$

18. $\dfrac{13}{20}$

19. $\dfrac{5}{8}$

20. $\dfrac{3}{16}$

21. $\dfrac{3}{40}$

22. $\dfrac{2}{5}$

23. $\dfrac{7}{16}$

24. $\dfrac{5}{32}$

8.3 Homework Cont.

Find the decimal equivalents rounded to the indicated place.

25. $\dfrac{4}{9}$ hundredth

26. $\dfrac{7}{9}$ hundredth

27. $\dfrac{7}{12}$ thousandth

28. $\dfrac{5}{12}$ thousandth

29. $\dfrac{4}{11}$ thousandth

30. $\dfrac{1}{3}$ hundredth

Find the decimal equivalents, using the bar notation.

31. $\dfrac{5}{6}$

32. $\dfrac{2}{9}$

33. $\dfrac{3}{11}$

34. $\dfrac{1}{3}$

35. $\dfrac{7}{11}$

36. $\dfrac{5}{18}$

8.3 Homework Cont.

Find the decimal equivalents for each of the following mixed numbers.

37. $5\dfrac{3}{4}$

38. $13\dfrac{3}{4}$

39. $12\dfrac{1}{2}$

40. $913\dfrac{3}{8}$

41. $17\dfrac{1}{4}$

42. $6\dfrac{5}{16}$

Change the following fractions to decimals, do not round off. (Use a repeating bar if necessary):

43. $21\dfrac{1}{3}$

44. $8\dfrac{7}{11}$

45. $143\dfrac{7}{12}$

46. $4\dfrac{11}{18}$

47. $39\dfrac{8}{27}$

48. $77\dfrac{5}{27}$

8.4 Comparing and Arranging with Decimals

Inequality Symbols

Recall the inequality symbols: $<$ and $>$.

For any numbers a and b:

1. $a < b$ (read a is less than b) means a is to the left of b on the number line.

2. $a > b$ (read a is greater than b) means a is to the right of b on the number line.

Note: The inequality symbol must always **point toward the smaller number**. You could also say it **opens up to the larger number**.

Examples: $16 > 5$ $3 < 8$

Decimals are compared in the same way we compared whole numbers: by comparing the place values from left to right.

Comparing Decimal Numbers

If the whole numbers to the left of the decimal point are different:

First, compare the whole numbers to the left of the decimal point.

If the numbers are not the same; the larger decimal number is the one with the larger whole number, the smaller decimal number is the one with the smaller whole number.

Example 1. Write the appropriate inequality symbol between the following numbers.

   a) 89.758   403.8   **b)** 67.25   63.48

Answer: For a), the larger number would be the decimal number to the right.

Looking at the whole numbers to the left of the decimal point, 403 is larger than 89.

For b), the larger number would be the decimal number to the left.

Looking at the whole numbers to the left of the decimal point, 67 is larger than 63.

   a) $89.758 < 403.8$ **b)** $67.25 > 63.48$

Your Turn Problem #1

Place the correct inequality symbol between the two numbers.

a)   412.98   73.216   **b)**   32.995   34.72

If the whole numbers to the left of the decimal point are the same:

If the whole numbers to the left of the decimal point are equal, we then compare the digits to the right of the decimal point to determine which is larger or smaller.

We compare the numbers by starting with the first place value, the tenths place. The number which has a larger digit in the tenths place is the larger number. If the digits of each number in the tenths place are also equal, we then move on to the next place value on the right, the hundredths place. Again, the number which has a larger digit in the hundredths place is the larger number. If the digits are again equal, we then move on to the thousandths place to determine which is larger. We continue this process until the larger number is identified.

Note: If a number does not have a digit in place value, then the digit is zero.

For example, 6.4 is the same as 6.40, and 7 is the same as 7.0.

Example 2. Write the appropriate inequality symbol between the following numbers.

       a)   5.7   5.8   **b)**   0.43   0.45

Answer: For a), the larger number would be the decimal number to the right.

           Looking at the tenths place, 8 is larger than 7.

           For b), the larger number would be the decimal number to the right.

           Looking at hundredths place, 5 is larger than 3.

       a) $\boxed{5.7 < 5.8}$ **b)** $\boxed{0.43 < 0.45}$

Your Turn Problem #2

Place the correct inequality symbol between the two numbers.

a)   2.7   2.5   **b)**   0.427   0.423

Example 3. Write the appropriate inequality symbol between the following numbers.

      a)   2.53   2.7   **b)**   0.4   0.28

Answer: For a), the larger number would be the decimal number to the right.

                  Looking at the tenths place, 7 is larger than 5. At first glance, some students make the mistake of thinking 2.53 is larger because 53 is larger than 7. However, if you write a zero in the hundredths place, 2.7 = 2.70. Now, it may be clearer that 2.70 is larger than 2.53.

                  For b), the larger number would be the decimal number to the left.

                  Looking at tenths place, 4 is larger than 2.

                  a) $\boxed{2.53 < 2.7}$ **b)** $\boxed{0.4 > 0.28}$

Your Turn Problem #3

Place the correct inequality symbol between the two numbers.

a)   7.8   7.25   **b)**   0.56   0.528

Example 4. Write the appropriate inequality symbol between the following numbers.

      a)   0.29573   0.2958   **b)**   0.3 $0.\overline{3}$

Answer: For a), the larger number would be the decimal number to the right.

                  Looking at the ten-thousandths place (4th decimal place), 8 is larger than 7.

                  For b), the larger number would be the decimal number to the right since

                  $0.\overline{3} = 0.3333...$

                  a) $\boxed{0.29573 < 0.2958}$ **b)** $\boxed{0.3 < 0.\overline{3}}$

Your Turn Problem #4

Place the correct inequality symbol between the two numbers.

a)   0.5374   0.53729   **b)** $0.\overline{4}$ 0.4

Comparing a fraction and a decimal

Fractions and decimals may be compared by first converting any fractions to decimals.

Example 5. Write the appropriate symbol ($<$, $>$, or $=$) between the following numbers.

a) $\dfrac{5}{8}$ 0.63   b) $\dfrac{4}{9}$ 0.4   c) $\dfrac{3}{5}$ 0.6

Answer: a) $\dfrac{5}{8} = 0.625$ b) $\dfrac{4}{9} = 0.44...$ c) $\dfrac{3}{5} = 0.6$

$0.625 \ < \ 0.63$ $0.44... > \ 0.4$ $0.6 \ = \ 0.6$

$$\boxed{\dfrac{5}{8} < 0.63}$$ $$\boxed{\dfrac{4}{9} > 0.4}$$ $$\boxed{\dfrac{3}{5} = 0.6}$$

Your Turn Problem #5

Write the appropriate symbol ($<$, $>$, or $=$) between the following numbers.

a) $\dfrac{1}{3}$ 0.3   b) $\dfrac{5}{16}$ 0.313   c) $5\dfrac{3}{4}$ 5.75

Arranging Decimals from Least to Greatest

Comparing decimals will serve us in arranging decimals from least to greatest. First, identify the smallest value by using the same techniques we have previously learned. Then from the remaining numbers, identify the next smallest number. Continue this process until the last number remaining is the largest number.

Example 6. Arrange the following decimals from least to greatest.

5.7, 2.5, 3.2, 4.1, 3.9

Answer: Looking at the whole numbers in front of the decimal point, 2.5 is the smallest since it has the smallest whole number. The next smallest whole number in front of the decimal point is 3. There are two numbers with a whole number of 3, which are 3.2 and 3.9. The smaller of these two numbers is 3.2 because the tenths place is smaller. After those two numbers 4.1 is the next smallest number and the largest number is 5.7

$$\boxed{2.5, \ 3.2, \ 3.9, \ 4.1 \ 5.7}$$

Your Turn Problem #6

Arrange the following decimals from least to greatest.

    1.9, 2.7, 1.3, 2.2, 3.4

Example 7. Arrange the following decimals from least to greatest.

    0.37, 0.2, 0.27, 0.5, 0.329

Answer: Looking at the tenths place, the two smallest numbers are 0.2 and 0.27. Comparing the hundredths place, 0.2 is the smallest. The next smallest decimal looking at the tenths place are 0.37 and 0.329. 0.329 is smaller because it has a smaller digit in the hundredths place. The largest number is 0.5 because it has the largest digit in the tenths place.

$$\boxed{0.2,\ \ 0.27,\ \ 0.329,\ \ 0.37,\ \ 0.5}$$

Your Turn Problem #7

Arrange the following decimals from least to greatest.

    0.435, 0.472, 0.44, 0.4, 0.401

If fractions appear along with decimals where we are to arrange the numbers in order from least to greatest, convert the fractions to decimals for comparison.

Example 8. Convert $\dfrac{5}{8}$ to a decimal.

Solution:

1. Divide the denominator into the numerator.
2. Insert a decimal point after the 5. Bring the decimal point straight up in the quotient.
3. Divide. Keep going until it ends or there is a repeating pattern.

Answer: $\boxed{0.625}$

$$
\begin{array}{r}
0.625 \\
8\overline{)5.000} \\
-48 \\
\hline
20 \\
-16 \\
\hline
40 \\
-40 \\
\hline
0
\end{array}
$$

Your Turn Problem #8

Convert $\dfrac{5}{12}$ to a decimal. Use a repeating decimal bar if necessary.

Example 9. Arrange the following decimals from least to greatest.

$$0.38, \quad \frac{1}{3}, \quad 0.4, \quad 0.3, \quad \frac{3}{8}$$

Answer: First, convert the fractions to decimals.

$$0.38, \quad 0.\overline{3}, \quad 0.4, \quad 0.3, \quad 0.375$$

Looking at the numbers with a three in the tenths place, the smallest number is 0.3.

Comparing the hundredths place, $0.\overline{3}$ is the next number because it has a 3 in the hundredths place. 0.375 is the next smallest decimal looking at the hundredths place is 0.38. 0.4 is the largest number because it has the largest digit in the tenths place. When writing the answer, use the original numbers given in the problem.

$$\boxed{0.3, \quad \frac{1}{3}, \quad \frac{3}{8}, \quad 0.38, \quad 0.4}$$

Your Turn Problem #9

Arrange the following decimals from least to greatest.

$$0.78, \quad \frac{7}{9}, \quad 0.8, \quad 0.7, \quad \frac{3}{4}$$

8.4 Homework: Comparing and Arranging with Decimals

Write the appropriate symbol (< , > , or =) between the following numbers.

1. 7.4   9.7   **2.** 34.82   28.9   **3.** 402.58   99.46

4. 37   24.98   **5.** 17.4   17.9   **6.** 3.58   3.27

7. 5.25   5.28   **8.** 0.473   0.472   **9.** 2.039   2.052

10. 45.27   45.4   **11.** 18.3   18.199   **12.** 2.058   2.102

13. 8.7384   8.739   **14.** 0.29812   0.29808   **15.** 4.225   4.2237

16. $\dfrac{3}{8}$ 0.38   **17.** $\dfrac{5}{9}$ 0.5   **18.** $\dfrac{3}{7}$ 0.4283

19. 0.35 $\dfrac{7}{20}$ **20.** 0.4329 $\dfrac{7}{16}$ **21.** 17.36 $17\dfrac{2}{5}$

22. 0.71398 $\dfrac{5}{7}$ **23.** $\dfrac{22}{3}$ 7.3   **24.** $35\dfrac{1}{8}$ 35.125

8.4 Homework cont.

Arrange the following decimals from least to greatest.

25. 7.7, 5.5, 5.9, 7.3, 9.1

26. 3.4, 5.2, 5.7, 3.2, 4.3

27. 0.43, 0.08, 0.6, 0.429, 0.472

28. 0.73, 0.705, 0.71, 0.725, 0.717

29. 1.05, 1.5, 1.45, 1.445, 1.405

30. 5.18, 5.8, 5.28, 5.09, 5.78

31. 0.73, $\dfrac{3}{4}$, 0.77, 0.8, $\dfrac{7}{10}$

32. 0.43, $\dfrac{2}{5}$, 0.36, 0.405, $\dfrac{9}{20}$

33. $\dfrac{8}{9}$, 0.88, 0.9, 0.8, $\dfrac{7}{8}$

34. $\dfrac{5}{11}$, $\dfrac{4}{9}$, 0.45, 0.5, 0.405

Practice Test 8

Perform the following operations.

1. $38.8 \div 8$

2. $30.855 \div 15$

3. $2.912 \div 1.4$

4. $78.84 \div 0.9$

5. $1.2974 \div 0.013$

6. $0.0867 \div 0.017$

7. $86753.09 \div 100$

8. $4.96 \div 1000$

Perform the following operations. Round to the indicated place value.

9. $16 \div 0.7$ tenth

10. $9.5 \div 14$ hundredth

11. $27 \div 0.11$ thousandth

12. $7.25 \div 1.3$ hundredth

Practice Test 8 cont.

Solve the following application problems.

13. Emilie purchased a 50 inch Sony LED television for \$936.96. If she pays it off in 12 equal payments, what amount will she be paying each month?

14. Usain Bolt ran the 100 meter three times at the 2008 Beijing Olympics. His times for each were as follows: 9.8 seconds, 9.69 seconds, and 9.74 seconds. What was his average time (to the nearest 100th) for the three sprints?

Find the area of the circles with the given information.

Use $A = \pi \times r^2$ **where** $\pi = 3.14$.

15. radius = 7 ft

16. diameter = 18 ft

Find the circumference of the circles with the given information.

Use $C = \pi \times d$ **and** $\pi = 3.14$.

17. diameter = 15 ft

18. radius = 11 ft

Practice Test 8 cont.

Write each of the following as a proper fraction or a mixed number. Reduce if possible.

19. 0.45

20. 7.25

21. 8.08

22. 15.122

Find the decimal equivalent for each of the following fractions.

23. $\dfrac{2}{5}$

24. $\dfrac{7}{8}$

25. $\dfrac{5}{16}$

26. $\dfrac{9}{40}$

Find the decimal equivalents rounded to the indicated place.

27. $\dfrac{5}{9}$ hundredth

28. $\dfrac{3}{7}$ thousandth

Practice Test 8 cont.

Find the decimal equivalents, using the bar notation.

29. $\dfrac{7}{11}$ **30.** $\dfrac{4}{15}$

Find the decimal equivalents for each of the following mixed numbers.

31. $8\dfrac{3}{5}$ **32.** $29\dfrac{5}{8}$

Change the following fractions to decimals, do not round off. (Use a repeating bar if necessary):

33. $185\dfrac{7}{15}$ **34.** $45\dfrac{5}{18}$

Write the appropriate symbol (< , > , or =) between the following numbers.

35. 9.2   2.93   **36.** 5.2736   5.27354   **37.** 0.123 $\dfrac{1}{8}$

Arrange the following decimals from least to greatest.

38. 0.34, 0.04, 0.3, 0.33, 0.335   **39.** 0.88, $\dfrac{4}{5}$, 0.85, 0.9, $\dfrac{7}{8}$

Practice Final

Rewrite the following using words.

1. 9,243,807: _____

Name the property that is illustrated

2. $9 + 0 = 9$: _____

3. $7 \cdot (2 \cdot 9) = (7 \cdot 2) \cdot 9$: _____

4. $9 + 2 = 2 + 9$: _____

5. $17 \cdot 1 = 17$: _____

6. $8 \cdot 0 = 0$: _____

7. $8 \cdot (5 + 7) = (8 \cdot 5) + (8 \cdot 7)$: _____

Round to the indicated place value.

8. 29,488 ; thousand

9. 5,954 ; hundred

Practice Final cont.

Perform the indicated operation.

10. $79+658$

11. $7,258+9,659+767$

12. $23,771-15,894$

13. $35,000-12,653$

14. 207×54

15. Find the product of 88 and 22.

16. What is the sum of 356 and 644?

17. Find the difference of 562 and 268.

Practice Final cont.

Perform the indicated division if possible.

18. $13 \div 0$

19. $0 \div 28$

20. $16,272 \div 8$

21. $9,243 \div 17$

Evaluate the following expressions.

22. 5^3

23. 9^0

24. 15^1

25. 1^{15}

Evaluate the following expressions.

26. $5 \times (36 \div 9 \times 4)$

27. $6^2 - (7 - 2)^2$

Practice Final cont.

Evaluate the following expressions.

28. $3 + 2(12 - 9)$

29. $2^5 \div (3^2 - 1)$

Solve the following applications.

30. Charles agreed to buy a new car paying $3200 down and $430 a month for 5 years. What is the total cost of the car?

31. There are 632 students who are taking a field trip. If each bus can hold 40 students, how many buses will be needed for the field trip?

32. A scooter holds 11 gallons of gas. If the scooter gets 78 miles per gallon, how far can the scooter go on a full tank of gas?

Practice Final cont.

Solve the following applications.

33. Alexis has 5 math exams with scores of 65, 72, 93, 91, and 94. Find the average of all five exams.

34. Monique has $800 to spend on school expenses. She spends $363 for tuition and $355 for books. How much money does she have left to spend on other school expenses?

35. Find the perimeter of the rectangle if the width is 9 ft and the length is 18 ft.

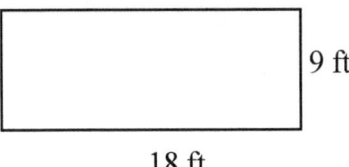

9 ft

18 ft

36. Find the area of the rectangle if the width is 5 m and the length is 22 m.

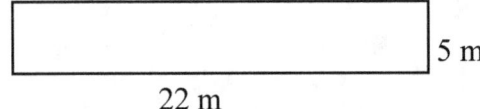

5 m

22 m

Practice Final cont.

Solve the following application.

37. Find the area of the triangle if the base is 35 ft and the height is 22 ft.

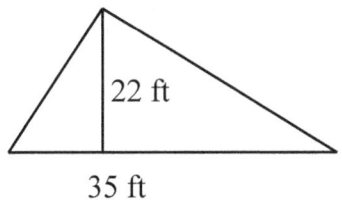

38. List the prime factorization of 126.

39. List all the factors of 40.

40. Find the GCF of 45 and 60.

41. Find the GCF of 44 and 66.

42. Find the LCM of 28 and 36.

43. Find the LCM of 14 and 18.

Practice Final cont.

Simplify the following if possible.

44. $\dfrac{60}{84}$

45. $\dfrac{8}{0}$

46. $\dfrac{0}{4}$

47. $\dfrac{34}{102}$

Perform the indicated operation. Do not leave improper fractions for answers.

48. $\dfrac{24}{35} \cdot \dfrac{21}{40}$

49. $\dfrac{7}{18} \cdot 24$

50. $4\dfrac{1}{5} \cdot \dfrac{9}{14}$

51. $4\dfrac{3}{8} \cdot 1\dfrac{1}{15}$

52. $\dfrac{22}{25} \div \dfrac{33}{35}$

53. $\dfrac{9}{13} \div 18$

54. $4\dfrac{2}{3} \div 4\dfrac{1}{5}$

55. $24 \div 5\dfrac{1}{3}$

Practice Final cont.

Perform the indicated operation. Do not leave improper fractions for answers.

56. $\dfrac{5}{21} + \dfrac{3}{14}$

57. $\dfrac{7}{8} + \dfrac{2}{5} + \dfrac{3}{4}$

58. $\dfrac{15}{16} - \dfrac{5}{12}$

59. $\dfrac{17}{30} - \dfrac{3}{10}$

60. $75\dfrac{7}{9} + 35\dfrac{5}{6}$

61. $18\dfrac{7}{15} + 12\dfrac{4}{5} + 10\dfrac{5}{6}$

62. $78\dfrac{3}{8} - 27\dfrac{2}{3}$

63. $18 - 4\dfrac{5}{16}$

Solve the following applications. Do not leave improper fractions for answers.

64. A roll of cable contains 40 yards. If $15\dfrac{3}{4}$ yards is cut from the roll, how much cable remains?

Practice Final cont.

Solve the following applications. Do not leave improper fractions for answers.

65. Find the area of a rectangle if the width is $2\frac{1}{3}$ ft and the length is $5\frac{1}{2}$ ft .

66. A certain size bottle holds exactly $\frac{2}{3}$ pint of liquid. How many of these bottles can be filled from a 24-pint container?

67. Mike has 100 ft of cable. He wants to cut it into pieces $3\frac{3}{4}$ ft long. How many pieces that are $3\frac{3}{4}$ ft long of cable will he have?

68. Find the perimeter of a rectangle if each width is $\frac{1}{4}$ yd and each length is $\frac{2}{3}$ yd.

69. Jimmy lives $\frac{3}{4}$ of a mile from school. After he walks $\frac{2}{3}$ mile, how much further is it to school?

Practice Final cont.

Solve the following applications. Do not leave improper fractions for answers.

70. Sandra spends $\frac{3}{8}$ of her monthly income on her car payment. If her monthly income is $864,

how much is her car payment?

71. In Bismarck, they received $9\frac{5}{16}$ inches of snow in December, $14\frac{2}{3}$ inches of snow in

January, and $14\frac{3}{8}$ inches of snow in February. What is the total amount of snowfall for the

three months?

Solve the following Equations. Do not leave improper fractions for answers.

72. $18 \cdot x = 15$

73. $\frac{5}{8} \cdot n = 40$

74. $\frac{10}{17} \cdot x = \frac{15}{34}$

75. $1\frac{4}{5} \cdot x = 18$

Practice Final cont.

Write in Words.

76. 53.048

Write in decimal form.

77. Five hundred ninety and thirty-eight thousandths

Round to the indicated place value:

78. 719.4635 ; hundredths

Perform the indicated operation.

79. $37 + 5.27 + 0.768$

80. $70 - 2.72$

81. 80.5×0.59

82. 3.6×1000

83. $13.452 \div 1.9$

84. $45.79 \div 100$

Practice Final cont.

Perform the following operations. Round to the indicated place value.

85. $17.3 \div 0.7$ thousandth

86. $35 \div 0.11$ tenths

Solve the following application problems.

87. Find the perimeter of a rectangle with a length of 15.8 meters and a width of 9.4 meters.

88. Trina's monthly pay check was for $3,532.65. The mortgage was $1,075.26, and the utilities added up to $157.34. How much does she have left?

Practice Final cont.

Solve the following application problems.

89. Fuji apples at a market cost $1.39 per pound. How much would 3.2 lbs cost? Round to the nearest cent.

90. Derric purchased a 50 inch Sony LED television for $1,055.16. If he pays it off in 12 equal payments, what amount will he be paying each month?

Write as a mixed number. Reduce if possible.

91. 9.28

Find the decimal equivalent for each fraction or mixed number. Round where indicated.

92. $\dfrac{3}{8}$

93. $17\dfrac{3}{4}$

94. $\dfrac{4}{7}$ thousandth

Find the decimal equivalent for the fraction using the repeating decimal bar.

95. $\dfrac{11}{15}$

Answers to Essential Mathematics Workbook

1.1 Answers: Place Value of Whole Numbers

YT#1 a) 7 ten thousands   b) 7 tens   c) c. 7 billions

YT#2 a) nine thousand, two hundred seven

b) three million, four hundred twenty-nine thousand, seven hundred eighteen

YT#3 a) 135,402   b) 4,000,408,046

1. hundreds   **2.** hundred thousands   **3.** hundred millions   **4.** ten billions

5. thousands   **6.** tens   **7.** twenty-one   **8.** nine hundred forty-one

9. three thousand, five hundred one   **10.** ninety-three thousand, eight hundred eighty

11. thirty-four million, fifty-eight thousand, twelve

12. seven trillion, twelve billion, thirty thousand, eight

13. 47   **14.** 327   **15.** 5,082   **16.** 135,402   **17.** 8,000,000,007,000

18. 2,400,011   **19.** 43,904   **20.** 8,000,280

1.2 Answers: Rounding and Ordering of Whole Numbers

YT#1	YT#2	YT#3	YT#4	YT#5
a) 30	a) 73,000	a) 45,000	a) 250,000	a) >
b) 860	b) 408,300	b) 0	b) 3,440,000	b) >
c) 6,800	c) 0	c) 1,000		c) <

1. 350	**2.** 3,510	**3.** 13,520	**4.** 2,400	**5.** 0
6. 440	**7.** 76,430	**8.** 571,600	**9.** 111,300	**10.** 700
11. 3,600	**12.** 14,000	**13.** 30,000	**14.** 700	**15.** 39,700
16. 711,900	**17.** 43,000	**18.** 100	**19.** 7,000	**20.** 18,000
21. 20,000	**22.** 0	**23.** 1,000	**24.** 3,000	**25.** 773,413,000
26. 10,000	**27.** 7,001,000	**28.** 180,000	**29.** 45,900,000	**30.** 227,000,000

31. 340,000,000   **32.** <   **33.** >   **34.** <   **35.** >   **36.** >   **37.** <

1.3 Answers: Addition of Whole Numbers

YT#1 a) Additive Identity Property   b) Commutative Property of Addition

c) Associative Property of Addition

YT#2 15,811   **YT#3** $65,500

1. Commutative Property of Addition   2. Associative Property of Addition

3. Associative Property of Addition   4. Additive Identity Property

5. Associative Property of Addition   6. Commutative Property of Addition

7. 17	**8.** 26	**9.** 39	**10.** 60	**11.** 113
12. 143	**13.** 79	**14.** 182	**15.** 477	**16.** 1,603
17. 806	**18.** 1,622	**19.** 1,814	**20.** 6,417	**21.** 139,251
22. 2,448	**23.** 158,450	**24.** 640,070	**25.** 6,207	**26.** 57,559
27. 40	**28.** 6,176	**29.** 172	**30.** 716	**31.** 556
32. 2,460 miles	**33.** $982			

1.4 Answers: Subtraction of Whole Numbers

YT#1 a) 34   b) 3   **YT#2** a) 258   b) 15678   c) 80,173   **YT#3** 11,092   **YT#4** 377

1. 13	**2.** 73	**3.** 28	**4.** 58	**5.** 6
6. 38	**7.** 14	**8.** 19	**9.** 35	**10.** 44
11. 419	**12.** 217	**13.** 46	**14.** 28	**15.** 27
16. 244	**17.** 187	**18.** 258	**19.** 363	**20.** 108
21. 1,746	**22.** 2,683	**23.** 5,663	**24.** 5,227	**25.** 67
26. 649	**27.** 7,590	**28.** 1,259	**29.** 5,785	**30.** 422
31. 555	**32.** 1,463	**33.** 74,395	**34.** 3	**35.** 10
36. 38	**37.** 352	**38.** 646	**39.** 556	**40.** 24
41. 13,846	**42.** 7	**43.** 52	**44.** $2,125	**45.** $513
46. $341	**47.** 83 points			

Practice Test 1 Answers

1. nine million, seven hundred forty-two thousand, thirteen

2. five billion, twenty-three million, nine hundred sixty thousand, two hundred twenty-seven

3. additive identity property   4. associative property of addition

5. commutative property of addition

6. 954	7. 1,377	8. 16,867	9. 88,212	10. 30,488
11. 1,024	12. 1,472	13. 760	14. 44,000	15. 50,000
16. 9,400	17. 200	18. 784	19. 30,974	20. 21,235
21. 4,725	22. 2,477	23. 14,589	24. 255	25. 336
26. <	27. >	28. >	29. <	

30. 1,834 more students   31. $108 left   32. 734 pages

2.1 Answers: Multiplication of Whole Numbers

YT #1   a)   Multiplicative Identity Property   b)   Commutative Property of Multiplication

     c)   Distributive Property   d)   Multiplication Property of Zero

     e)   Associative Property of Multiplication

YT #2 a) 216   b) 2,082   **YT#3** a) 1,260   b) 20,288

YT#4 a) 420   b) 13,320   **YT#5** a) 63,000   b) 1,400,000

1. Commutative Property of Multiplication   2. Multiplicative Identity Property

3. Multiplication Property of Zero   4. Commutative Property of Multiplication

5. Associative Property of Multiplication   6. Multiplication Property of Zero

7. Distributive Property   8. Multiplicative Identity Property

9. Associative Property of Multiplication

10. 96	11. 105	12. 192	13. 252	14. 1,541
15. 2,448	16. 6,080	17. 1,872	18. 1,748	19. 682
20. 1,768	21. 8,304	22. 11,736	23. 40,638	24. 26,208
25. 22,248	26. 263,160	27. 94,500	28. 561,176	29. 177,600
30. 609,364	31. 1,431,565	32. 1,409,382	33. 1,040	34. 267,716
35. 1,437,696	36. 1,500,000	37. 720,000	38. 280,000	39. 8,400,000
40. 84	41. 21,504	42. 375 cal.	43. $1,890	44. 408 seats
45. 60 apt.	46. 14,000 koi	47. 432 computers	48. 240 miles	

2.2 Answers: Division of Whole Numbers

YT#1 a) 21   b) 34   c) 85   d) 7   **YT#2** a) 1   b) 1   c) 1   d) 1

YT#3 a) 0   b) 0   c) 0   d) 0

YT#4 a) undefined   b) undefined   c) undefined   d) undefined

YT#5 a) indeterminate   b) indeterminate   **YT#6** 532   **YT#7** 70 R 5

YT#8 a) 15   b) 7   c) 18   **YT#9** a) 1   b) 32   c) 1

YT#10 a) undefined   b) undefined   c) 0   d) indeterminate

1. undefined	2. 0	3. indeterminate	4. 1	5. 1
6. 47	7. undefined	8. 0	9. 147	10. 254
11. 367	12. 543	13. undefined	14. 0	15. 307
16. 9,703	17. indeterminate	18. undefined	19. 23	20. 54
21. 84	22. 673 r 3	23. 15,322 r 1	24. 706	25. 20 r 8
26. 3,465	27. 2,318	28. 31 r 11	29. 2,181 r 1	30. 8,075 r 12
31. undefined	32. 0	33. 68 r 20	34. 10,678 r 7	35. 51 r 6
36. 33	37. 49 r 5	38. 127	39. 18 r 12	40. 5 r 6
41. indeterminate	42. undefined	43. 13 buses	44. 32 sections	
45. 34 mpg	46. 107 minutes	47. 367 bottles	48. 12	49. 24
50. 1	51. 1	52. 0	53. 0	54. 16
55. undefined	56. 0	57. indeterminate	58. 1	59. 28

2.3 Answers: Exponents

YT#1 a) 125   b) 49   c) 27   d) 16   **YT#2** a) 4   b) 76

YT 3 a) 1   b) 1   **YT#4** 189   **YT#5** 3,375   **YT#6** 100,000

YT#7 9,000   **YT#8** 2,000,000

1. 16	2. 27	3. 100	4. 125	5. 1
6. 23	7. 343	8. 144	9. 0	10. 1
11. 216	12. 243	13. 1	14. 9	15. 1
16. 1	17. 64	18. 10,000	19. 49	20. 128
21. 1	22. 1	23. 64	24. 1	25. 91
26. 7	27. 54	28. 2	29. 0	30. 5
31. 576	32. 2	33. 16	34. 64	35. 432
36. 8	37. 120,000	38. 32,000	39. 32,400,000	40. 0
41. 12	42. 101	43. 2	44. 0	

2.4 Answers: Order of Operations

YT#1 a) 12 b) 20 c) 36   **YT#2** a) 16   b) 0   **YT#3** a) 27   b) 14

1. 14	**2.** 16	**3.** 11	**4.** 14	**5.** 12	**6.** 8
7. 3	**8.** 16	**9.** 4	**10.** 9	**11.** 81	**12.** 125
13. 128	**14.** 25	**15.** 96	**16.** 17	**17.** 81	**18.** 64
19. 6	**20.** 18	**21.** 1	**22.** 3	**23.** 11	**24.** 17
25. 9	**26.** 8	**27.** 13	**28.** 40	**29.** 50	**30.** 28
31. 29	**32.** 31	**33.** 2	**34.** 15	**35.** 120	**36.** 63
37. 150	**38.** 73	**39.** 32	**40.** 14		

Practice Test 2 Answers

1. Commutative Property of Multiplication   **2.** Multiplicative Identity Property

3. Multiplication Property of Zero   **4.** Associative Property of Multiplication

5. Distributive Property   **6.** 243   **7.** 1,776   **8.** 31,995

9. 2,613   **10.** undefined   **11.** 0   **12.** 1   **13.** 13

14. 246   **15.** 5024   **16.** 359 R 19   **17.** 33 R 6   **18.** 209 R 6

19. 259 R 8   **20.** 81   **21.** 1   **22.** 9   **23.** 1

24. 8   **25.** 48   **26.** 126   **27.** 180   **28.** 16

29. 36   **30.** 4   **31.** 36   **32.** 8   **33.** 8

34. 20   **35.** 73

3.1 Answers: Solving Equations

YT#1 a) $N = 15,000$ b) $N = 705$ **YT#2** a) $x = 566$ b) $x = 47$ c) $A = 26$

YT#3 a) 18   b) 1   c) 0   **YT#4** a) $x = 48$ b) $x = 203$

YT#5 a) $x = 32$ b) $x = 205$ **1.** $x = 14$ **2.** $x = 14$

3. $N = 3$ **4.** $N = 36$ **5.** $x = 27$ **6.** $a = 22$

7. $x = 187$ **8.** $x = 11$ **9.** $N = 456$ **10.** $Y = 176$

11. $x = 39$ **12.** $c = 50$ **13.** $x = 63$ **14.** $x = 1,655$

15. $A = 13$ **16.** $B = 48$ **17.** $x = 84$ **18.** $b = 1$

19. $x = 10$ **20.** $x = 72$ **21.** $C = 17$ **22.** $x = 96$

23. $x = 23$ **24.** $b = 56$ **25.** $N = 76,000$ **26.** $H = 50$

27. $x = 7$ **28.** $x = 25$ **29.** $x = 37$ **30.** $x = 86$

3.2 Answers: Word Problems using Whole Numbers

YT#1 $65,500 **YT#2** 11,092 **YT#3** 476 miles

YT#4 568 people **YT#5** 86 **YT#6** 102°

YT#7 He has enough money. He will get back $4. **YT#8** 18 runs

1. $982	**2.** $704	**3.** $341	**4.** 1316 seats
5. 43 mpg	**6.** $345	**7.** 2,400 labels	**8.** $15,600
9. $13,800	**10.** 14 buses	**11.** 108 sections	**12.** $32,350
13. 81	**14.** $768	**15.** 78	**16.** $33,832

17. 62 cases **18.** She has enough. She will get back $43. **19.** 106 bones

3.3 Answers: Geometric Applications - Perimeter

YT#1 20 **YT#2** 12 ft **YT#3** 6 **YT#4** 7, 6, and 19

YT#5 16 and 10 **YT#6** 62 ft **YT#7** 88 ft

1. 16 ft	**2.** 20 yd	**3.** 19 ft	**4.** 46 ft	**5.** 76 ft
6. 42 ft	**7.** 66 ft	**8.** 58 m	**9.** 172 ft	**10.** 68 ft
11. 44 ft	**12.** 288 ft	**13.** 346 yd	**14.** 56 ft	**15.** 76 ft

3.4 Answers: Geometric Applications – Area and Volume

YT#1 154 m^2 **YT#2** 77 m^2 **YT#3** 243 m^2 **YT#4** 193 ft^2 **YT#5** 328 ft^2

YT#6 945 ft^3 **1.** 81 ft^2 **2.** 84 yd^2 **3.** 60 ft^2 **4.** 54 km^2

5. 165 ft^2 **6.** 180 mi^2 **7.** 624 ft^2 **8.** 520 in^2 **9.** 27 ft^3

10. 300 m^3 **11.** 270 ft^2 **12.** 4,700 ft^2 **13.** 323 ft^2 **14.** $3,600

15. $3,136 **16.** $3,168

3.5 Answers: Divisibility Rules

YT#1   a) Yes. 90 ends with an even digit.

        b) No. 243 ends with an odd digit.

        c) Yes. 2,518 ends with an even digit.

YT#2   a) Yes. $1 + 7 + 1 = 9$ and 9 is divisible by 3.

        b) Yes. $2 + 9 + 7 = 18$ and 18 is divisible by 3.

        c) No. $6 + 1 + 3 = 10$ and 10 is not divisible by 3.

YT#3   a) Yes. 230 ends with a 0 or a 5.

        b) Yes. 175 ends with a 0 or a 5.

        c) No. 352 does not end with a 0 or a 5.

YT#4   a) Yes. $5 + 4 + 8 + 1 = 18$ and 18 is divisible by 9.

        b) No. $3 + 2 + 9 = 14$ and 14 is not divisible by 9.

        c) Yes. $5 + 0 + 4 = 9$ and 9 is divisible by 9.

YT#5   a) Yes. 540 ends with a 0.

        b) Yes. 300 ends with a 0.

        c) No. 255 does not end with a 0.

YT#6   a) Yes. 7 divides evenly into 77. $7 \times 11 = 77$.

        b) Yes. 7 divides evenly into 161. $7 \times 23 = 161$.

        c) No. 7 does not divide evenly into 131.

YT#7   710 is divisible by 2, 5 and 10.

1. 2, 3, 5, 10   **2.** 2   **3.** 3   **4.** 2, 5, 10   **5.** 3

6. 2, 3   **7.** 2, 3   **8.** none   **9.** 2, 3, 5, 10   **10.** 3

11. yes, $7 \cdot 13 = 91$ **12.** no   **13.** yes, $7 \cdot 19 = 133$ **14.** yes, $7 \cdot 11 = 77$

15. yes, $7 \cdot 209 = 1,463$ **16.** 72, 158, 260, 378, 570, 4,530, 8,300

17. 45, 72, 378, 570, 585, 4,530   **18.** 45, 260, 570, 585, 4,530, 8,300

19. 260, 570, 4,530, 8,300   **20.** 45, 72, 378, 585

3.6 Answers: Prime and Composite Numbers

YT#1. composite, divisible by 3   **YT#2.** composite, divisible by 5   **YT#3** prime

1. 2, 3, 5, 7, 11, 13, 17, 19, 23, 29, 31, 37, 41, 43, 47, 53, 59, 61, 67, 71, 73, 79, 83, 89, 97

2. prime   **3.** composite, divisible by 3   **4.** prime   **5.** composite, divisible by 3

6. composite, divisible by 2 (and 3)   **7.** prime   **8.** composite, divisible by 7

9. composite, divisible by 3   **10.** composite, divisible by 5

11. composite, divisible by 2   **12.** 9, 18, 27, 36, 45, 54, 63, 72, 81, 90

13. 25, 50, 75, 100, 125, 150, 175, 200, 225, 250   **14.** 11, 22, 33, 44, 55, 66, 77, 88, 99, 110

3.7 Answers: Factors and Prime Factorization

YT#1. $2 \cdot 2 \cdot 3 \cdot 5$ or $2^2 \cdot 3 \cdot 5$ **YT#2.** 1, 2, 3, 6, 9, 18

YT#3. 1, 2, 4, 5, 8, 10, 16, 20, 40, 80

1. $2^2 \cdot 3^2$ **2.** $2^3 \cdot 3$ **3.** $2^3 \cdot 3^2 \cdot 5$ **4.** $2^4 \cdot 3$

5. $2^2 \cdot 5^2$ **6.** $5 \cdot 13$ **7.** 13, prime number   **8.** $5 \cdot 11$

9. $2^4 \cdot 3 \cdot 5$ **10.** $2^4 \cdot 3^2$ **11.** $3 \cdot 7^2$ **12.** $5^2 \cdot 7$

13. $3 \cdot 5 \cdot 13$ **14.** $2 \cdot 7 \cdot 11$ **15.** $5^3 \cdot 7$ **16.** $2^3 \cdot 3^2 \cdot 7$

17. 1, 2, 3, 4, 6, 9, 12, 18, 36   **18.** 1, 2, 3, 4, 6, 8, 12, 24

19. 1, 3, 5, 15   **20.** 1, 2, 3, 4, 6, 8, 12, 16, 24, 48

21. 1, 2, 4, 5, 10, 20, 25, 50, 100   **22.** 1, 5, 13, 65

23. 1, 13   **24.** 1, 5, 11, 55

25. 1, 2, 4, 8, 16, 32, 64   **26.** 1, 2, 3, 6

27. 1, 2, 3, 5, 6, 9, 10, 15, 18, 30, 45, 90   **28.** 1, 2, 3, 4, 5, 6, 10, 12, 15, 20, 30, 60

29. 1, 2, 3, 5, 6, 10, 15, 30   **30.** 1, 2, 4, 7, 8, 14, 28, 56

3.8 Answers: The GCF (Greatest Common Factor)

YT#1 14   **YT#2** 44   **YT#3** 1   **YT#4** 13

1. 6   **2.** 5   **3.** 2   **4.** 10   **5.** 7   **6.** 12   **7.** 13

8. 11   **9.** 25   **10.** 1   **11.** 46   **12.** 19

3.9 Answers: The LCM (Least Common Multiple)

YT #1 630   **YT#2** 1,200   **YT#3** 275   **YT#4** 230

1. 45   **2.** 216   **3.** 315   **4.** 2,280   **5.** 72

6. 40   **7.** 80   **8.** 120   **9.** 525   **10.** 72

11. 72   **12.** 1,260   **13.** 315   **14.** 240   **15.** 36

16. 60   **17.** 120   **18.** 216   **19.** 165   **20.** 72

Practice Test 3 Answers

1. $23   **2.** $18,600   **3.** 2,580 pages   **4.** 11 buses

5. 228 miles   **6.** $48,880   **7.** 81   **8.** 38 ft

9. $48\,m^2$ **10.** $300\,ft^2$ **11.** $144\,ft^2$

12. Prime. 59 is only divisible by 1 and itself.   **13.** Not prime. 231 is divisible by 3.

14. $2^4 \cdot 3$ **15.** $2^4 \cdot 3 \cdot 5$ **16.** $3 \cdot 5^2 \cdot 7$ **17.** 1, 2, 3, 5, 6, 10, 15, 30

18. 1, 3, 5, 9, 15, 45   **19.** 12   **20.** 14   **21.** 140

22. 120   **23.** 126   **24.** 180

4.1 Answers: Introduction to Fractions

YT#1 $\dfrac{7}{10}$

YT#2 a) $\dfrac{5}{12}$ **b)** $\dfrac{7}{16}$ **c)** $\dfrac{17}{20}$

YT#3 a) 29 **b)** 29 **c)** 13

YT#4 a) 1 **b)** 1 **c)** 1

YT#5 a) 8 **b)** 1 **c)** 356

YT#6 a) $\dfrac{1}{12}$ **b)** $\dfrac{1}{4}$ **c)** $\dfrac{1}{153}$

YT#7 a) 0 **b)** 0 **c)** 0

YT#8 a) undefined **b)** undefined **c)** undefined

YT#9 indeterminate

YT#10 a) $2\dfrac{4}{7}$ **b)** $8\dfrac{3}{4}$ **c)** $5\dfrac{2}{11}$

YT#11 a) $\dfrac{19}{2}$ **b)** $\dfrac{51}{4}$ **c)** $\dfrac{53}{3}$

1. numerator = 6, denominator = 13
2. $\dfrac{3}{5}$
3. $\dfrac{1}{4}$
4. $\dfrac{7}{12}$
5. $\dfrac{11}{17}$

6. $\dfrac{6}{17}$
7. $\dfrac{5}{12}$
8. $\dfrac{16}{35}$
9. $\dfrac{2}{9}$
10. $\dfrac{5}{12}$
11. $\dfrac{9}{70}$
12. $\dfrac{5}{21}$

13. proper fraction
14. improper fraction
15. mixed number
16. 1

17. undefined
18. proper fraction
19. mixed number

20. improper fraction
21. improper fraction
22. improper fraction

23. proper fraction
24. 4
25. 45
26. 1
27. 19
28. 23

29. 58
30. 35
31. 19
32. 11
33. undefined
34. 0

35. 1
36. 17
37. undefined
38. $1\dfrac{2}{5}$
39. $2\dfrac{4}{9}$
40. $4\dfrac{4}{5}$

41. $6\dfrac{1}{7}$
42. $4\dfrac{2}{9}$
43. $5\dfrac{2}{11}$
44. $12\dfrac{1}{6}$
45. $2\dfrac{6}{17}$
46. $4\dfrac{13}{25}$

47. $3\dfrac{23}{24}$
48. $15\dfrac{2}{5}$
49. $8\dfrac{8}{11}$
50. $\dfrac{22}{5}$
51. $\dfrac{44}{7}$
52. $\dfrac{19}{2}$

53. $\dfrac{367}{3}$
54. $\dfrac{89}{5}$
55. $\dfrac{3}{2}$
56. $\dfrac{100}{3}$
57. $\dfrac{200}{9}$
58. $\dfrac{50}{3}$
59. $\dfrac{163}{2}$

4.2 Answers: Simplifying Fractions

YT#1 a) $\frac{3}{4}$ **b)** $\frac{11}{15}$ **YT#2 a)** $\frac{5}{6}$ **b)** $\frac{14}{15}$ **YT#3 a)** $\frac{1}{6}$ **b)** 6 **c)** $\frac{1}{4}$

YT#4 $3\frac{3}{5}$ **YT#5** $3\frac{3}{8}$ **YT#6** $\frac{3}{13}$

1. $\frac{3}{4}$ 2. $\frac{11}{15}$ 3. $\frac{4}{5}$ 4. $\frac{2}{3}$ 5. $\frac{9}{25}$

6. $\frac{4}{7}$ 7. $\frac{3}{4}$ 8. $\frac{1}{2}$ 9. $\frac{15}{22}$ 10. $\frac{1}{6}$

11. 0   12. 55   13. $\frac{3}{5}$ 14. $\frac{11}{15}$ 15. $\frac{22}{45}$

16. $\frac{3}{4}$ 17. $\frac{2}{3}$ 18. $\frac{2}{5}$ 19. 1   20. undefined

21. $\frac{4}{5}$ 22. $\frac{2}{3}$ 23. $\frac{29}{37}$ 24. $\frac{19}{25}$ 25. $\frac{5}{7}$

26. $\frac{7}{8}$ 27. $\frac{7}{13}$ 28. $\frac{3}{5}$ 29. $\frac{3}{5}$ 30. $\frac{7}{8}$

31. $\frac{2}{3}$ 32. $\frac{29}{129}$ 33. $\frac{7}{8}$ 34. $\frac{7}{8}$ 35. $\frac{1}{2}$

36. $\frac{1}{3}$ 37. $2\frac{1}{5}$ 38. $2\frac{2}{3}$ 39. $3\frac{3}{4}$ 40. $3\frac{1}{2}$

41. $2\frac{1}{11}$ 42. $5\frac{1}{3}$ 43. $8\frac{1}{3}$ 44. $3\frac{1}{3}$ 45. $1\frac{4}{5}$

46. $1\frac{1}{7}$ 47. $5\frac{3}{5}$ 48. $2\frac{16}{17}$ 49. $\frac{3}{20}$ 50. $\frac{4}{5}$

51. $\frac{1}{5}$ 52. $\frac{7}{8}$ 53. $\frac{4}{5}$

4.3 Answers: Reading a Ruler

1. $1\frac{1}{2}$ in   2. $1\frac{3}{16}$ in   3. $\frac{1}{2}$ in   4. $1\frac{1}{8}$ in   5. $\frac{3}{8}$ in

6. $2\frac{1}{4}$ in   7. $2\frac{3}{4}$ in

Practice Test 4 Answers

1. $3^3 \cdot 5$　　　**2.** $2^2 \cdot 3 \cdot 7^2$　　　**3.** 1, 3, 5, 9, 15, 45

4. 1, 2, 3, 4, 6, 8, 9, 12, 18, 24, 36, 72　　　　　**5.** 140　　　**6.** 182

7. Proper fractions: $\dfrac{8}{9}$, $\dfrac{9}{16}$　Improper fractions: $\dfrac{7}{3}$, $\dfrac{18}{5}$, $\dfrac{15}{15}$, Mixed Numbers: $3\dfrac{1}{5}$, $23\dfrac{5}{9}$

8. 23　　　**9.** undefined　　　**10.** 37　　　**11.** 0　　　**12.** $\dfrac{5}{7}$

13. $\dfrac{1}{5}$　　**14.** $\dfrac{2}{3}$　　**15.** $\dfrac{17}{28}$　　**16.** $\dfrac{3}{5}$　　**17.** $\dfrac{1}{5}$

18. $\dfrac{5}{9}$　　**19.** $3\dfrac{2}{5}$　　**20.** $5\dfrac{3}{11}$　　**21.** $8\dfrac{2}{9}$　　**22.** $3\dfrac{3}{4}$

23. $5\dfrac{2}{3}$　　**24.** $4\dfrac{1}{6}$　　**25.** $\dfrac{29}{9}$　　**26.** $\dfrac{87}{5}$　　**27.** $\dfrac{152}{3}$

28. $\dfrac{3}{4}''$　　**29.** $1\dfrac{5}{8}''$

5.1 Answers: Multiplication with Fractions

YT#1 a) $\dfrac{14}{33}$ **b)** $\dfrac{12}{25}$　　　**YT#2 a)** $\dfrac{1}{6}$ **b)** $\dfrac{1}{9}$　　　**YT#3 a)** 4　**b)** 14

YT#4 a) $3\dfrac{3}{4}$ **b)** 15　　　**YT#5 a)** $18\dfrac{3}{4}$ **b)** 8　　　**YT#6 a)** $\dfrac{3}{20}$ **b)** $3\dfrac{3}{4}$

YT#7 $\dfrac{1}{4}$　　　**YT#8** 160　　　**YT#9** 20ft^2　　　**YT#10** $6\dfrac{1}{2}\text{ft}^2$

YT#11 $33\dfrac{3}{4}\text{m}^2$　　**YT#12** $800

1. $\dfrac{9}{25}$　　**2.** $\dfrac{9}{20}$　　**3.** $\dfrac{1}{10}$　　**4.** $\dfrac{3}{10}$　　**5.** $2\dfrac{2}{3}$

6. $\dfrac{18}{35}$　　**7.** $\dfrac{4}{9}$　　**8.** 1　　**9.** $3\dfrac{1}{3}$　　**10.** $8\dfrac{1}{6}$

11. $4\dfrac{2}{3}$　　**12.** $41\dfrac{1}{3}$　　**13.** $5\dfrac{1}{19}$　　**14.** $16\dfrac{30}{49}$　　**15.** $\dfrac{15}{128}$

16. $\dfrac{3}{35}$　　**17.** $2\dfrac{1}{2}$　　**18.** $\dfrac{3}{8}$　　**19.** $\dfrac{4}{39}$　　**20.** $7\dfrac{1}{5}$

5.1 Answers: Multiplication with Fractions cont.

21. 1　　**22.** $\dfrac{1}{216}$　　**23.** $\dfrac{7}{12}$　　**24.** $\dfrac{15}{44}$　　**25.** $\dfrac{3}{10}$

26. $10\dfrac{1}{2}$　　**27.** $\dfrac{27}{80}$　　**28.** $85\dfrac{3}{4}$　　**29.** $178\dfrac{1}{2}$　　**30.** 2,448

31. 2,415　　**32.** $1\dfrac{4}{27}$　　**33.** $8\dfrac{1}{6}$　　**34.** 2,520　　**35.** 21 ft

36. 24 women　　**37.** \$740　　**38.** $13\dfrac{7}{8}$ ft^2　　**39.** $27\dfrac{17}{48}$ ft^2　　**40.** $2\dfrac{1}{4}$ mi

41. $3\dfrac{3}{4}$ cups　　**42.** $12\dfrac{1}{12}$ yd^3　　**43.** 8 ft^2　　**44.** 42 ft^2　　**45.** $14\dfrac{7}{8}$ ft^2

5.2 Answers: Division with Fractions

YT#1 $\dfrac{8}{15}$　　**YT#2** $\dfrac{4}{9}$　　**YT#3** 11 pieces　　**YT#4** \$63,000

1. 1　　**2.** $\dfrac{9}{20}$　　**3.** $1\dfrac{5}{6}$　　**4.** $\dfrac{16}{27}$　　**5.** 2

6. $\dfrac{1}{16}$　　**7.** $1\dfrac{31}{65}$　　**8.** $\dfrac{35}{54}$　　**9.** $\dfrac{4}{5}$　　**10.** 27

11. $1\dfrac{5}{9}$　　**12.** $\dfrac{15}{64}$　　**13.** $\dfrac{2}{7}$　　**14.** $1\dfrac{17}{18}$　　**15.** $\dfrac{1}{10}$

16. 1　　**17.** $1\dfrac{1}{9}$　　**18.** $2\dfrac{2}{9}$　　**19.** $\dfrac{2}{3}$　　**20.** 2

21. $6\dfrac{3}{4}$　　**22.** $\dfrac{5}{9}$　　**23.** $1\dfrac{1}{4}$　　**24.** $\dfrac{7}{18}$　　**25.** 26 shirts

26. 25 bottles　　**27.** 11 pieces　　**28.** 16 pieces　　**29.** \$34,000　　**30.** $2\dfrac{7}{8}$ ml

5.3 Answers: Solving Equations of the Form a · x = b

YT#1 45

YT#2 $x = 8\frac{2}{11}$

YT#3 $x = \frac{5}{8}$

YT#4

$a = 64$

YT#5 $y = \frac{14}{15}$

YT#6 $x = \frac{2}{33}$

YT#7 $x = 1\frac{5}{28}$

1. $x = 77$

2. $a = 59$

3. $x = 208$

4. $c = 147$

5. $h = 3\frac{7}{8}$

6. $x = 4\frac{2}{5}$

7. $x = 8\frac{1}{5}$

8. $x = 2\frac{1}{2}$

9. $x = \frac{2}{5}$

10. $x = \frac{7}{13}$

11. $d = \frac{1}{3}$

12. $x = \frac{1}{4}$

13. $y = 18$

14. $n = 45$

15. $x = 17\frac{3}{5}$

16. $x = 31\frac{1}{2}$

17. $x = 2\frac{2}{5}$

18. $x = 1\frac{7}{20}$

19. $m = 2\frac{1}{4}$

20. $x = \frac{1}{15}$

21. $x = \frac{3}{44}$

22. $x = 15$

23. $g = 2\frac{2}{9}$

24. $x = 1\frac{1}{4}$

Practice Test 5 Answers

1. $\frac{10}{21}$

2. $5\frac{1}{4}$

3. 9

4. 1

5. $4\frac{1}{8}$

6. $3\frac{1}{2}$

7. $\frac{20}{21}$

8. $\frac{3}{4}$

9. $\frac{1}{24}$

10. $2\frac{1}{4}$

11. $4\frac{1}{2}$

12. $8\frac{4}{13}$

13. 435 students

14. $420

15. $3\frac{3}{8}$ mi

16. $\frac{3}{8}$ of a cup

17. $19\frac{5}{6}$ ft^2

18. 80 bottles

19. 60 shirts

20. 35 pieces

21. $x = 2\frac{1}{2}$

22. $x = \frac{1}{3}$

23. $n = 81$

24. $x = 1\frac{1}{20}$

25. $x = 7\frac{1}{2}$

26. $x = \frac{2}{5}$

6.1 Answers: Reviewing the LCM and Equivalent Fractions

YT#1 140   **YT#2** $\dfrac{28}{48}$ **YT#3** $\dfrac{153}{969}$ **YT#4** $\dfrac{70}{5}$ **YT#5** $\dfrac{11}{38}$

1. 60   **2.** 24   **3.** 84   **4.** 84   **5.** 55   **6.** 12

7. 78   **8.** 102   **9.** 45   **10.** 300   **11.** 105   **12.** 1260

13. $\dfrac{24}{32}$ **14.** $\dfrac{15}{51}$ **15.** $\dfrac{64}{16}$ **16.** $\dfrac{125}{25}$ **17.** $\dfrac{56}{8}$ **18.** $\dfrac{54}{6}$

19. $\dfrac{60}{64}$ **20.** $\dfrac{33}{54}$ **21.** $\dfrac{42}{98}$ **22.** $\dfrac{15}{42}$ **23.** $\dfrac{49}{168}$ **24.** $\dfrac{65}{169}$

25. $\dfrac{159}{424}$ **26.** $\dfrac{153}{272}$ **27.** $\dfrac{340}{800}$ **28.** $\dfrac{176}{550}$ **29.** $\dfrac{4}{17}$ **30.** $\dfrac{11}{40}$

6.2 Answers: Addition with Fractions

YT#1 $\dfrac{7}{10}$ **YT#2** $\dfrac{37}{90}$ **YT#3** $\dfrac{23}{40}$ **YT#4** $\dfrac{163}{180}$ in

1. $\dfrac{4}{5}$ **2.** $\dfrac{1}{3}$ **3.** 1   **4.** $\dfrac{2}{3}$ **5.** $1\dfrac{1}{2}$

6. 1   **7.** $1\dfrac{2}{13}$ **8.** $1\dfrac{11}{12}$ **9.** $\dfrac{11}{12}$ **10.** $\dfrac{13}{24}$

11. $\dfrac{19}{24}$ **12.** $\dfrac{31}{36}$ **13.** $\dfrac{29}{60}$ **14.** $\dfrac{17}{24}$ **15.** $\dfrac{27}{50}$

16. $1\dfrac{5}{12}$ **17.** $\dfrac{19}{24}$ **18.** $\dfrac{13}{30}$ **19.** $\dfrac{173}{1260}$ **20.** $1\dfrac{7}{180}$

21. $\dfrac{61}{66}$ **22.** $\dfrac{29}{45}$ **23.** $\dfrac{31}{40}$ **24.** $\dfrac{47}{72}$ **25.** $\dfrac{83}{120}$

26. $1\dfrac{5}{72}$ **27.** $\dfrac{1264}{1575}$ **28.** $\dfrac{559}{630}$ **29.** $2\dfrac{11}{18}$ yd   **30.** $1\dfrac{1}{8}$ in

31. $\dfrac{17}{30}$ **32.** $\dfrac{103}{120}$ ft

6.3 Answers: Subtracting and Comparing Fractions

YT#1 $\frac{1}{2}$ **YT#2** $\frac{13}{90}$ **YT#3** $\frac{37}{120}$ **YT#4** $1\frac{11}{90}$ **YT#5** $\frac{7}{18}<\frac{9}{20}$ **YT#6** $\frac{11}{18},\frac{2}{3},\frac{7}{9}$

1. $\frac{2}{5}$ 2. $\frac{1}{5}$ 3. $\frac{2}{5}$ 4. $\frac{3}{8}$ 5. $\frac{19}{36}$

6. $\frac{13}{48}$ 7. $\frac{11}{24}$ 8. $\frac{7}{30}$ 9. $\frac{277}{1260}$ 10. $\frac{67}{180}$

11. $\frac{29}{66}$ 12. $\frac{4}{9}$ 13. $\frac{19}{40}$ 14. $\frac{41}{72}$ 15. $\frac{37}{120}$

16. $\frac{31}{72}$ 17. $\frac{416}{1575}$ 18. $\frac{269}{630}$ 19. $\frac{133}{180}$ 20. $\frac{19}{30}$

21. $\frac{13}{24}$mi 22. $\frac{31}{84}$ 23. $\frac{1}{45}$ 24. $\frac{1}{16}$lb 25. $\frac{13}{24}$gal

26. $\frac{1}{8}$mi 27. $\frac{2}{5}$in 28. > 29. < 30. >

31. > 32. > 33. > 34. $\frac{7}{20},\frac{3}{8},\frac{5}{12}$ 35. $\frac{9}{16},\frac{7}{12},\frac{2}{3}$

36. $\frac{17}{45},\frac{13}{30},\frac{12}{25}$ 37. $\frac{1}{5},\frac{5}{24},\frac{9}{40}$

6.4 Answers: Addition with Mixed Numbers

YT#1 $8\frac{2}{9}$ **YT#2** $151\frac{11}{20}$ **YT#3** $346\frac{23}{60}$ **YT#4** $85\frac{11}{14}$ **YT#5** $24\frac{11}{20}$

YT#6 $32\frac{11}{12}$ft 1. $17\frac{1}{2}$ 2. $28\frac{1}{2}$ 3. $11\frac{2}{3}$ 4. $87\frac{1}{2}$

5. $9\frac{13}{24}$ 6. $17\frac{1}{2}$ 7. $15\frac{23}{36}$ 8. $12\frac{7}{12}$ 9. $20\frac{3}{10}$

10. $18\frac{13}{48}$ 11. $37\frac{11}{18}$ 12. $442\frac{13}{30}$ 13. $103\frac{3}{22}$ 14. $38\frac{31}{72}$

15. $91\frac{31}{60}$ 16. $22\frac{23}{48}$ 17. $6\frac{13}{20}$ 18. $19\frac{1}{12}$ 19. $17\frac{7}{8}$ 20. $27\frac{11}{12}$

21. $18\frac{4}{9}$ 22. 58 23. $41\frac{1}{4}$ft 24. $13\frac{1}{2}$in 25. $88\frac{1}{6}$ft 26. $41\frac{19}{24}$ft

6.5 Answers: Subtraction with Mixed Numbers

YT#1 $\frac{23}{18}$ **YT#2** $34\frac{11}{60}$ **YT#3** $52\frac{7}{12}$ **YT#4** $16\frac{5}{12}$ **YT#5** $20\frac{8}{15}$

YT#6 $21\frac{5}{8}$ **YT#7** $13\frac{5}{12}$ **YT#8** $11\frac{7}{12}$ **YT#9** $5\frac{8}{15}$ hr   **1.** $3\frac{1}{3}$

2. $5\frac{1}{12}$ **3.** $8\frac{11}{18}$ **4.** $8\frac{5}{12}$ **5.** $2\frac{5}{6}$ **6.** $11\frac{43}{60}$

7. $5\frac{23}{48}$ **8.** $5\frac{7}{18}$ **9.** $7\frac{7}{12}$ **10.** $7\frac{13}{24}$ **11.** $4\frac{19}{20}$

12. $11\frac{9}{20}$ **13.** $6\frac{1}{8}$ **14.** $1\frac{1}{6}$ **15.** $4\frac{3}{4}$ **16.** $5\frac{3}{5}$

17. $3\frac{5}{9}$ **18.** $15\frac{19}{40}$ **19.** $4\frac{8}{9}$ **20.** $8\frac{3}{4}$ **21.** $10\frac{1}{5}$

22. $14\frac{25}{36}$ **23.** $\frac{7}{12}$ lb   **24.** $13\frac{1}{6}$ yd   **25.** $25\frac{3}{4}$ ft   **26.** $7\frac{17}{24}$ hr   **27.** $2\frac{7}{12}$ lbs

---

6.6 Answers: Order of Operations with Fractions

YT#1 a) 12 **b)** 20   **YT#2 a)** 9 **b)** 56   **YT#3** $\frac{7}{9}$ **YT#4** $2\frac{1}{2}$ **YT#5** $\frac{1}{9}$

1. $\frac{1}{2}$ **2.** $11\frac{3}{4}$ **3.** $1\frac{1}{10}$ **4.** $2\frac{5}{6}$ **5.** $12\frac{4}{5}$ **6.** $15\frac{1}{10}$

7. $\frac{27}{64}$ **8.** $11\frac{14}{25}$ **9.** $\frac{41}{72}$ **10.** $3\frac{6}{19}$ **11.** $\frac{81}{175}$ **12.** 1

Practice Test 6 Answers

1. 165   **2.** 420   **3.** $\frac{1}{2}$ **4.** $\frac{22}{45}$ **5.** $1\frac{7}{30}$

6. $1\frac{19}{144}$ **7.** $\frac{1}{9}$ **8.** $\frac{1}{9}$ **9.** $\frac{19}{60}$ **10.** $\frac{23}{42}$

11. $35\frac{1}{3}$ **12.** $158\frac{31}{40}$ **13.** $45\frac{3}{10}$ **14.** $18\frac{3}{20}$ **15.** $26\frac{11}{12}$

16. $29\frac{23}{24}$ **17.** $60\frac{1}{3}$ **18.** $7\frac{7}{15}$ **19.** $15\frac{33}{80}$ **20.** $12\frac{8}{13}$

21. $1\frac{5}{6}$ yd   **22.** 26 ft   **23.** $\frac{3}{8}$ mi   **24.** $\frac{11}{15}$ **25.** $41\frac{5}{12}$ in

26. $\frac{11}{16}$ lb   **27.** $31\frac{1}{3}$ yd   **28.** $57\frac{7}{24}$ ft   **29.** $15\frac{31}{60}$ hr

7.1 Answers: Introduction to Decimals

YT#1 a) seventy-five hundredths b) ninety-one and thirty-six thousandths

YT#2 Fifty-eight and nine hundred forty-three thousandths

YT#3 a) 0.7 b) 0.13 c) 67.015

YT#4 a) $\dfrac{2}{5}$ b) $\dfrac{13}{25}$ c) $12\dfrac{3}{50}$

YT#5 a) 9.6 b) 205.8 c) 0.8

YT#6 a) 7.25 b) 72.00 c) 0.78

YT#7 a) 5.286 b) 0.0713 c) 20

1. eight tenths 2. twenty-five hundredths 3. forty-four thousandths

4. twelve and three tenths 5. fifteen and eighty-seven hundredths

6. forty-three and thirty-four thousandths 7. 0.8 8. 0.24

9. 0.036 10. 12.4 11. 9.054 12. 0.007 13. 5.122

14. 0.043 15. 1.732 16. 2.002 17. 0.567 18. $\dfrac{2}{5}$

19. $\dfrac{1}{4}$ 20. $\dfrac{91}{1000}$ 21. $3\dfrac{4}{5}$ 22. $5\dfrac{21}{100}$ 23. $16\dfrac{3}{1000}$

24. $27\dfrac{1}{8}$ 25. 0.3 26. 0.29 27. 0.037 28. 12.08

29. 34.017 30. 79.3 31. 0.5 32. 0.3 33. 0.1

34. 3.9 35. 6.0 36. 16.1 37. 0.47 38. 0.25

39. 0.92 40. 3.88 41. 5.99 42. 16.07 43. 0.468

44. 0.254 45. 0.916 46. 3.883 47. 5.995 48. 16.068

49. 0.60 50. 2.9299 51. 16.0 52. 7.287 53. 0.04

54. 10 55. 314.91 56. 79.6120 57. 0

7.2 Answers: Adding and Subtracting Decimals

YT#1 17.93 **YT#2** 4.25 **YT#3** $15.52

1. 19.92 2. 22.25 3. 7.56 4. 3.39 5. $621.57

6. 15.901 7. 6.32 8. 21.45 9. 1.73 10. 4.43

11. 12.28 12. 2.57 13. 12.21 14. 3.846 15. $9.68

16. 69.4 in 17. $1296.06 18. 44.4 gal. 19. $175.75 20. 18.13 mi.

21. 192.2 sec 22. $85.33 23. $25.01

7.3 Answers: Multiplying with Decimals

YT#1 0.06072   **YT#2** 0.4641   **YT#3** a) 84.3   **b)** 867.5309

1. 51.52	**2.** 19.47	**3.** 141.96	**4.** 53.75	**5.** 12.18
6. 1.8312	**7.** 0.00034	**8.** 0.05225	**9.** 155.2	**10.** 140.2
11. 250	**12.** 125.61	**13.** 5540	**14.** 43.2	**15.** 48,131
16. $2,574.72	**17.** $16.63	**18.** $18.90/hr	**19.** $655.20	

Practice Test 7 Answers

1. thirty-five thousandths   **2.** eighty-two and forty-one hundredths   **3.** 19.253

4. 1.032   **5.** $9\dfrac{2}{5}$ **6.** $5\dfrac{7}{20}$ **7.** 0.17   **8.** 44.019

9. 5.9	**10.** 16.9	**11.** 0.43	**12.** 12.50	**13.** 867.531
14. 3.1416	**15.** 18.26	**16.** $389.64	**17.** 9.843	**18.** 16.62
19. 15.64	**20.** $15.14	**21.** 534.48	**22.** 0.1092	**23.** 3.14
24. 5230	**25.** 46.6 meters	**26.** $1,875.75	**27.** 207.28 sec	**28.** $684.44
29. $54.73	**30.** $5,490.72	**31.** $3.43	**32.** $22.20	**33.** $658.60

8.1 Answers: Dividing with Decimals

YT#1 3.97   **YT#2** 51.39   **YT#3** 8.75   **YT#4** 7.55

YT#5 0.062   **YT#6** a) 0.052   b) 867.5309

1. 2.37	**2.** 0.235	**3.** 6.321	**4.** 1.251	**5.** 9.1
6. 0.254	**7.** 2.034	**8.** 5.214	**9.** 6.23	**10.** 0.023
11. 56.4	**12.** 19.402	**13.** 21.4	**14.** 23.8	**15.** 102.125
16. 5.125	**17.** 1.552	**18.** 1.402	**19.** 0.025	**20.** 12.561
21. 0.0554	**22.** 4.326	**23.** 0.048131	**24.** 0.00578	**25.** 4.94
26. 34.62	**27.** 6.2	**28.** 0.26	**29.** 77.143	**30.** 8.49
31. $12.41	**32.** $61.84	**33.** 20.8		

8.2 Answers: Circumference and Area of a Circle

YT#1 10.5 ft **YT#2** 36 ft **YT#3** 153.86 ft^2 **YT#4** 530.66 ft^2

YT#5 15.7 ft **YT#6** 38.936 ft

1. 254.34 ft^2 2. 452.16 ft^2 3. 38.465 mi^2 4. 200.96 ft^2

5. 176.625 ft^2 6. 346.185 in^2 7. 40.82 ft 8. 18.84 ft

9. 11.618 mi 10. 81.64 ft 11. 113.04 ft 12. 27.004 in

8.3 Answers: Fractions and Decimals

YT#1 a) $\dfrac{2}{5}$ b) $\dfrac{13}{25}$ c) $12\dfrac{3}{50}$ **YT#2** a) 0.7 b) 0.13 c) 67.015

YT#3 0.75 **YT#4** $0.\overline{63}$ **YT#5** 0.54 **YT#6** 31.125 **YT#7** $21.\overline{45}$

1. $\dfrac{2}{5}$ 2. $\dfrac{1}{4}$ 3. $\dfrac{91}{1000}$ 4. $3\dfrac{4}{5}$ 5. $5\dfrac{21}{100}$

6. $16\dfrac{3}{1000}$ 7. $27\dfrac{1}{8}$ 8 $71\dfrac{3}{5}$ 9. $23\dfrac{12}{25}$ 10. $101\dfrac{1}{20}$

11. $13\dfrac{1}{40}$ 12. $8\dfrac{3}{20}$ 13. 0.25 14. 0.75 15. 0.6

16. 0.9 17. 0.375 18. 0.65 19. 0.625 20. 0.1875

21. 0.075 22. 0.4 23. 0.4375 24. 0.15625 25. 0.44

26. 0.78 27. 0.583 28. 0.417 29. 0.364 30. 0.33

31. $0.8\overline{3}$ 32. $0.\overline{2}$ 33. $0.\overline{27}$ 34. $0.\overline{3}$ 35. $0.\overline{63}$

36. $0.2\overline{7}$ 37. 5.75 38. 13.75 39. 12.5 40. 913.375

41. 17.25 42. 6.3125 43. $21.\overline{3}$ 44. $8.\overline{63}$ 45. $143.58\overline{3}$

46. $4.6\overline{1}$ 47. $39.\overline{296}$ 48. $77.\overline{185}$

8.4 Answers: Comparing and Arranging with Decimals

YT#1 a) $412.98 > 73.216$ **b)** $32.995 < 34.72$ **YT#2 a)** $2.7 > 2.5$ **b)** $0.427 > 0.423$

YT#3 a) $7.8 > 7.25$ **b)** $0.56 > 0.528$ **YT#4 a)** $0.5374 > 0.53729$ **b)** $0.\overline{4} > 0.4$

YT#5 a) $\dfrac{1}{3} > 0.3$ **b)** $\dfrac{5}{16} < 0.313$ **c)** $5\dfrac{3}{4} = 5.75$

YT#6 1.3, 1.9, 2.2, 2.7, 3.4   **YT#7** 0.4, 0.401, 0.435, 0.44, 0.472

YT#8 $0.41\overline{6}$ **YT#9** 0.7, $\dfrac{3}{4}$, $\dfrac{7}{9}$, 0.78, 0.8

1. $7.4 < 9.7$ 2. $34.82 > 28.9$ 3. $402.58 > 99.46$

4. $37 > 24.98$ 5. $17.4 < 17.9$ 6. $3.58 > 3.27$

7. $5.25 < 5.28$ 8. $0.473 > 0.472$ 9. $2.039 < 2.052$

10. $45.27 < 45.4$ 11. $18.3 > 18.199$ 12. $2.058 < 2.102$

13. $8.7384 < 8.739$ 14. $0.29812 > 0.29808$ 15. $4.225 > 4.2237$

16. $\dfrac{3}{8} < 0.38$ 17. $\dfrac{5}{9} > 0.5$ 18. $\dfrac{3}{7} > 0.4283$

19. $0.35 = \dfrac{7}{20}$ 20. $0.4329 < \dfrac{7}{16}$ 21. $17.36 < 17\dfrac{2}{5}$

22. $0.71398 < \dfrac{5}{7}$ 23. $\dfrac{22}{3} > 7.3$ 24. $35\dfrac{1}{8} = 35.125$

25. 5.5, 5.9, 7.3, 7.7, 9.1   26. 3.2, 3.4, 4.3, 5.2, 5.7

27. 0.08, 0.429, 0.43, 0.472, 0.6   28. 0.705, 0.71, 0.717, 0.725, 0.73

29. 1.05, 1.405, 1.445, 1.45, 1.5   30. 5.09, 5.18, 5.28, 5.78, 5.8

31. $\dfrac{7}{10}$, 0.73, $\dfrac{3}{4}$, 0.77, 0.8,   32. 0.36, $\dfrac{2}{5}$, 0.405, 0.43, , $\dfrac{9}{20}$

33. 0.8, $\dfrac{7}{8}$, 0.88, $\dfrac{8}{9}$, 0.9   34. 0.405, $\dfrac{4}{9}$, 0.45, $\dfrac{5}{11}$, 0.5

Practice Test 8 Answers

1. 4.85
2. 2.057
3. 2.08
4. 87.6
5. 99.8

6. 5.1
7. 867.5309
8. 0.00496
9. 22.9
10. 0.68

11. 245.455
12. 5.58
13. $78.08
14. 9.74 sec
15. 153.86 ft^2

16. 254.34 ft^2
17. 47.1 ft
18. 69.08 ft
19. $\dfrac{9}{20}$
20. $7\dfrac{1}{4}$

21. $8\dfrac{2}{25}$
22. $15\dfrac{61}{500}$
23. 0.4
24. 0.875
25. 0.3125

26. 0.225
27. 0.56
28. 0.429
29. $0.\overline{63}$
30. $0.2\overline{6}$

31. 8.6
32. 29.625
33. $185.4\overline{6}$
34. $45.2\overline{7}$

35. 9.2 > 2.93
36. 5.2736 > 5.27354
37. $0.123 < \dfrac{1}{8}$

38. 0.04, 0.3, 0.33, 0.335, 0.34
39. $\dfrac{4}{5}$, 0.85, $\dfrac{7}{8}$, 0.88, 0.9

Practice Final Answers

1. nine million, two hundred forty-three thousand, eight hundred seven

2. additive identity
3. associative property of multiplication

4. commutative property of addition
5. multiplicative identity property

6. multiplication property of zero
7. distributive property

8. 29,000
9. 6,000
10. 737
11. 17,684
12. 7,877

13. 22,347
14. 11,178
15. 1,936
16. 1,000
17. 294

18. undefined
19. 0
20. 2034
21. 543 R 12
22. 125

23. 1
24. 15
25. 1
26. 80
27. 11

28. 9
29. 4
30. $29,000
31. 16 buses
32. 858 miles

33. 83
34. $82 left
35. 54 ft
36. $110 \, m^2$
37. $385 ft^2$

38. $2 \cdot 3^2 \cdot 7$
39. 1, 2, 4, 5, 8, 10, 20, 40
40. GCF = 15
41. GCF = 22

42. LCM = 252
43. LCM = 126
44. $\dfrac{5}{7}$
45. undefined
46. 0

47. $\dfrac{1}{3}$
48. $\dfrac{9}{25}$
49. $9\dfrac{1}{3}$
50. $2\dfrac{7}{10}$
51. $4\dfrac{2}{3}$

52. $\dfrac{14}{15}$
53. $\dfrac{1}{26}$
54. $1\dfrac{1}{9}$
55. $4\dfrac{1}{2}$
56. $\dfrac{19}{42}$

57. $2\dfrac{1}{40}$
58. $\dfrac{25}{48}$
59. $\dfrac{4}{15}$
60. $111\dfrac{11}{18}$
61. $42\dfrac{1}{10}$

62. $50\dfrac{17}{24}$
63. $13\dfrac{11}{16}$
64. $24\dfrac{1}{4}$ yd
65. $12\dfrac{5}{6}$ ft^2
66. 36 bottles

67. 26 pieces
68. $1\dfrac{5}{6}$ yd
69. $\dfrac{1}{12}$ mi
70. $324
71. $38\dfrac{17}{48}$ in

72. $x = \dfrac{5}{6}$
73. n = 64
74. $x = \dfrac{3}{4}$
75. $x = 10$

76. fifty-three and forty-eight thousandths
77. 590.038
78. 719.46
79. 43.038

80. 67.28
81. 47.495
82. 3600
83. 7.08
84. 0.4579

85. 24.714
86. 318.2
87. 50.4 m
88. $2,300.05
89. $4.45

90. $87.93
91. $9\dfrac{7}{25}$
92. 0.375
93. 17.75
94. 0.571

95. $0.7\overline{3}$